Nanophotonics and Micro/Nano Optics International Conference (NANOP 2017)

Journal of Physics: Conference Series Volume 961

Barcelona, Spain
13 - 15 September 2017

Editor:

Muhammad Faryad

ISBN: 978-1-5108-5757-5
ISSN: 1742-6588

Printed from e-media with permission by:

Curran Associates, Inc.
57 Morehouse Lane
Red Hook, NY 12571

Some format issues inherent in the e-media version may also appear in this print version.

Printed by Curran Associates, Inc. (2018)

For permission requests, please contact the Institute of Physics
at the address below.

Institute of Physics
Dirac House, Temple Back
Bristol BS1 6BE UK

Phone: 44 1 17 929 7481
Fax: 44 1 17 920 0979

techtracking@iop.org

Additional copies of this publication are available from:

Curran Associates, Inc.
57 Morehouse Lane
Red Hook, NY 12571 USA
Phone: 845-758-0400
Fax: 845-758-2633
Email: curran@proceedings.com
Web: www.proceedings.com

Table of contents

Volume 961

Nanophotonics and Micro/Nano Optics International Conference 2017

13–15 September 2017, Barcelona, Spain

Accepted papers received: 19 January 2018
Published online: 1 February 2018

Preface

Papers

IOP Conf. Series: Journal of Physics: Conf. Series **961** (2017) 011001 doi:10.1088/1742-6596/961/1/011001

Preface

Introduction:

Owing to the unique optical properties of nanometre-scale systems, Nanophotonics has become a very efficient tool to control light-matter interaction beyond the limit of diffraction. Recent developments in this field have shown the immense potential of Nanophotonics to contribute to a wide spectrum of scientific disciplines, ranging from fundamental physics to optical engineering, and from biology to clean energies.

This second edition of the NANOP conference focuses on the transversal nature of Nanophotonics, by offering an international forum on the latest advances in the field from basic to applied research.

NANOP 2017 was held in Barcelona and hosted 215 delegates from 40 different countries.

Group Picture:

Main Sponsor: HUBNER Photonics (https://www.hubner-photonics.com/)

Chairman:

- Prof. Romain Quidant, ICFO – The Institute of Photonic Sciences, Spain

Guest Editor Committee

- Dr. Muhammad Faryad, Department of Physics, Lahore University of Management Sciences, Pakistan

IOP Conf. Series: Journal of Physics: Conf. Series **961** (2017) 011001 doi:10.1088/1742-6596/961/1/011001

Reviewers Committee:

Pierre Gilliot, France
Dai-Sik Kim, Korea, Republic of
Zhimin Shi, United States
Youcef Ouerdane, France
Vladimir Drachev, United States
Vlad Stolojan, United Kingdom
Vincenzo Giannini, United Kingdom
Vincent Paillard, France
Viktor Myroshnychenko, Germany
Urcan Guler, United States
Takashi Yatsui, Japan
Taiichi Otsuji, Japan
Teresa Monteiro, Portugal
Safi Jradi, France
Pavel Zemanek, Czech Republic
Pablo Alonso-González,
Nikitas Papasimakis, United Kingdom
Nicolas Izard, France
Mircea Dragoman, Romania
Martina Gerken, Germany
Martin Siler, Czech Republic
Marco Felici, Italy
Marco Cannas, Italy
Makoto Naruse, Japan
Luiz Gustavo Cancado, Brazil
K. V. Adarsh, India
Kevin Peter O'Donnell, United Kingdom
Jérôme Tignon, France
Jean-Marc Fedeli, France
Jaroslav Hamrle, Czech Republic
Jussi Toppari, Finland
Godefroy Leménager, France
Giuseppe Leo, France
Ferruccio Pisanello, Italy
Fabrice Pardo, France
Fabrice P. Laussy,
Francois Dubin, France
Doo Jae Park, Korea, Republic of
David S Bradshaw, United Kingdom
Dangyuan Lei, Hong Kong
Delphine Marris-Morini, France
Charles Cornet, France
Beniamino Sciacca, Netherlands
Annamaria Gerardino, Italy
Anna Vinattieri, Italy
Andrey B. Evlyukhin, Germany
Andreas Tittl, Switzerland
André Strittmatter, Germany
Alejandro Manjavacas,
Alberto Bramati, France
Adam Schwartzberg, United States
Michael Wolfgang Geiselmann, Switzerland

IOP Conf. Series: Journal of Physics: Conf. Series **961** (2017) 011001 doi:10.1088/1742-6596/961/1/011001

Raineri Fabrice, France
Barbay Sylvain, France
Braive Rémy, France
Jérôme Plain, France
Aurélien Crut, France
Nicolas Bonod, France
Sebastien Bidault, France
Jerome Wenger, France
Fabrice Charra, France
Jean-Francois Roch, France
Yannick De Wilde, France
Hamza Kurt, Turkey
Vijaya Prakash Gaddam, India
Jalil Ali, Malaysia
Ryan Beams,
Anton Kuzyk, Finland
Humeyra Caglayan, Finland

IOP Conf. Series: Journal of Physics: Conf. Series **961** (2017) 011002 doi:10.1088/1742-6596/961/1/011002

Peer review statement

All papers published in this volume of *Journal of Physics: Conference Series* have been peer reviewed through processes administered by the proceedings Editors. Reviews were conducted by expert referees to the professional and scientific standards expected of a proceedings journal published by IOP Publishing.

Helical light emission from plasmonic vortices via magnetic tapered tip

N Maccaferri[1*], Y Gorodetski[2], and D Garoli[1]

[1]Istituto Italiano di Tecnologia, Via Morego 16, 16136 Genova, Italy

[2]Mechanical Engineering Department and Electrical Engineering Department, Ariel University, 40700 Ariel, Israel

*nicolo.maccaferri@iit.it

Abstract. We investigate an architecture where a plasmonic vortex excited in a gold surface propagates on an adiabatically tapered magnetic tip and detaches to the far-field while carrying a well-defined optical angular momentum. We analyze the out-coming light and show that, despite generally high losses of flat magnetic surface, our 3D structure exhibits high energy throughput. Moreover, we show that once a magneto-optical activity is activated inside the magnetic tip a modulation of the total power transmittance is possible.

1. Introduction

Recently, structured optical beams became a subject of an intense research [1], due to numerous potential applications they offer in the fields of super-resolution imaging [2], optical tweezing [3], nanomanipulation [4] and telecommunications [5]. A special interest is dedicated to investigation of the interaction of structured light with metallic nanostructures, resulting in Surface Plasmon Polaritons (SPPs) carrying angular momentum (AM) [6–13]. These surface confined electromagnetic distributions are generally defined by a field singularity surrounded by a helical phase front, referred here as plasmonic vortices (PVs). The singularity strength is termed the topological charge of a vortex, and defined by the phase ramp acquired in one round trip about the singularity center. This charge is proportional to the orbital AM (OAM) carried by the field [9]. PVs are usually generated by coupling propagating vortex beams to the plasmonic mode on a metal surface by using center-symmetric coupling structure, known also as plasmonic vortex lenses (PVLs) [9]. A PVL generally comprises of periodic spiral or circular grooves or slits milled in the metal [6–9, 12–18]. The total AM of the resulting PV carries contributions from both the incident OAM and the additional OAM introduced by the coupling structure. Some more complex architectures proposed both coupling of light into a PV and then coupling of the propagating PV to the far-field, generating a scattered beam carrying non-zero orbital AM [6,8,15,16,18,19]. In this case, the PV generated by interaction of the incident light with a PVL can be modulated in the near-field and then scattered into free space by a proper decoupling structure. Most of the proposed systems utilized scattering as a primary coupling mechanism [6,16,19]. Although, it was shown that the latter method provides a way to abruptly modify the mode's AM it demonstrates a very low transmittance efficiency and not pure polarization state [19]. Recently we have experimentally demonstrated an efficient PV coupling to the free space by means of a single-layer PVL structure via adiabatically tapered gold tip at its center [6]. We have shown that by properly shaping the nanotip geometry the PV excited at the spiral structure can be coupled to far-field distribution carrying well defined OAM. Nevertheless, these effects can be solely controlled by the properly designed 3D shape of the tip and by the illuminating beam parameters. Various practical applications in nanophotonics require fast external control of the emerging beam characteristics. In this regard magnetoplasmonic devices draw a very promising route to active nanophotonics, since an externally applied magnetic field can alter the plasmonic response leading to novel and unexpected effects [20–29]. Obviously the main requirement of these devices is to be fabricated from the material that has both good plasmonic and magnetic properties, for instance nickel or cobalt [22, 23]. Here we demonstrate an architecture where a plasmonic vortex excited by a PVL milled in a gold surface propagates on an adiabatically tapered magnetic tip and detaches to the far-field while carrying a well-defined OAM. We analyze the light transmittance and show that despite generally high losses of flat nickel surface our 3D structure exhibits relatively high energy thoughput providing optimistic forecast for active magnetically-driven photonics.

2. Numerical simulations

In Fig. 1(a) we depict a scheme of the proposed structure, which comprises a multiple-turn spiral slit milled on a 300 nm gold film deposited onto a 100 nm Si_3N_4 membrane and a conical metallic tip located exactly at the center of the spiral. In this work the tip is made of nickel and cobalt. We first assume zero magnetic field or magnetization within the magnetic tip, thus the dielectric constant of nickel and cobalt is given by a single complex scalar. In the last part we assume that a magnetic field is applied along the tip (z-direction) [30, 31]. Three main parameters define the tip shape, namely its height (h), aperture angle (α), and curvature radius at the basis (r_c). The curvature radius at the tip apex is kept fixed to 20 nm throughout the paper. It is worth mentioning that the fabrication technique developed in previous works [6] enables to faithfully reproduce the designed structure proposed here.

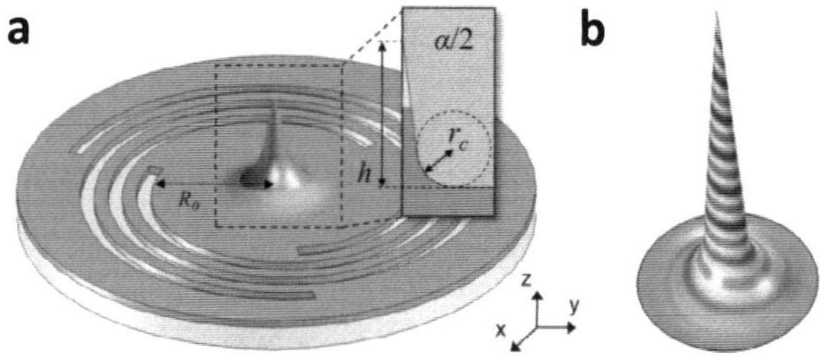

Fig. 1. (a) Scheme of the PVL. The geometrical parameters of the spiral slits are slit width = 200 nm, period= 763 nm, gold thickness = 100 nm, Si_3N_4 thickness = 100 nm. The nickel tip has h=6.5 μm; α=15° and r_c=1.57 μm. (b) Calculated z-component of the electric field propagating along the tip.

We analyze the case of a circularly polarized plane-wave impinging normally from the Si_3N_4 side. As has been described elsewhere [7–9,14], the electric field component locally orthogonal to the slits efficiently couples to the SPP mode of the metal-air interface. The plasmonic wavefronts launched by each spiral constructively interfere (the spiral period is λ_{SPP}) producing a PV radially propagating towards the PVL center. In the absence of the tip, the PV confined by the spiral grooves forms a standing wave, giving rise to the characteristic Bessel interference pattern [6]. The z-component of the electric field, in cylindrical coordinates, is proportional to $J_l(k_{SPP}r)exp(-\kappa z)$ $\exp(il\phi)$, where k_{SPP} is the wave vector of an SPP propagating on a flat gold-air interface, J_l is the lth order Bessel function of first kind and $\kappa=\sqrt{k_{SPP}^2-k_0^2}$, where $k_0=2\pi/\lambda_0$ is the wave vector of light in vacuum. It can be shown that the PV carries an orbital OAM quanta of $l\hbar$ where l is the PV topological charge, given by $l = m + s_i$ with the incident spin number $s_i = \pm 1$ corresponding to the right hand and left hand circularly polarized light, respectively, and m defining the topological charge of the spiral [7–9,14]. When the conical tip is present at the PVL center the surface confined electromagnetic mode may couple to the guided mode propagating along the tip upwards. In [6] we compared the behaviors of the PV with l = 1 excited on two structures having similar conical tips but with different curvature radius at the basis. We have shown that in the case of the negligible curvature radius ($r_c \rightarrow 0$) most of the energy is reflected back while a small fraction of the incoming energy is scattered to the far-field from the tip base. By increasing of the curvature radius at the tip basis (in our design r_c = 1.57 μm) the scattering diminishes and the mode is smoothly guided towards the top of the tip, and the spiral phase-fronts of the PV propagate on the tip almost without the losses (see the E_z component of the plasmonic field simulated by the COMSOL Multiphysics® software in Fig. 1(b)). The smoothed tip in this configuration perfectly matches between the PV and the corresponding plasmonic mode of the conical waveguide. This guided mode experiences gradual acceleration due to a local increasing of the effective index, and at some height, where its momentum matches the one of the free-space radiation, it detaches to the far-field. This has been demonstrated, both by means of finite element simulations and experimental measurements for a gold

PVL and tip [6]. Here we replace, the gold tip with a magnetic one, made of nickel or cobalt. The use of a magnetic material implies potentially interesting effects turning our metal structure into an optically active device by applying an external magnetic field. In the following part we compare the behavior of the previously investigated gold structure with a similar structure where the tip is completely, or partially made of a magnetic material.

Figure 2 reports the |E| maps for the PVs with $l = 0$, 1, 2 and 3 for a tip made of nickel. As can be seen, the behavior is similar with respect to the gold tip reported in [6]. For $l = 0$ the PV propagates to the end of the tip and is almost perfectly reflected back, which can be deduced by observing the interference pattern along the metal surface. This is the only mode that experiences an increasing of the effective index and slows down along the tip giving rise to the mode localization at the tip apex. For $l = 1$ our nickel tip nicely guides the PV to the tip apex where it beams out as a perfect Gaussian source. For higher l ($l > 1$), the PVs are efficiently decoupled to the free space in lower points and characteristic doughnut shaped distributions are expected in the far-field.

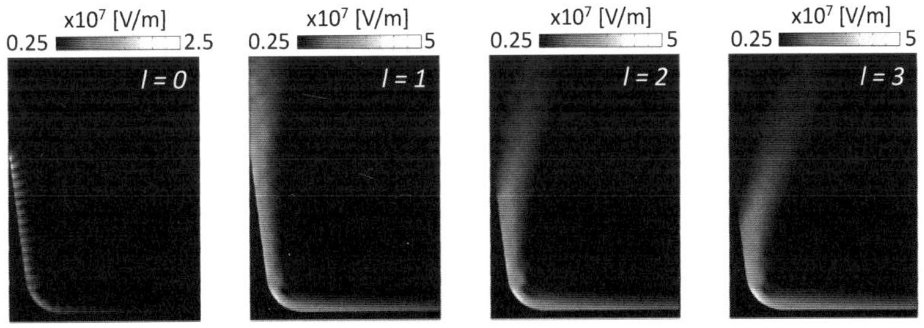

Fig. 2. |E| maps in case of a smoothed conical nickel tip illuminated by a PV with the indicated l values.

A smooth connection of the tip base insures almost lossless coupling of the plasmonic mode at the surface to the mode guided at the cone. The only losses in the system are attributed to the plasmonic absorption in the metal. By replacing the gold with nickel we introduce higher ohmic losses, however as can be seen in Fig. 2 the system still produces relatively efficient beaming. In Fig. 3(a) we summarize the PV total power transmittance as a function of l for various configurations comprising Au, Ni and Co tip. The transmittance efficiency spans from 10% for $l = 1$ PV and up to 35% at $l = 5$. In a comparison with the transmittance efficiency of the recently proposed holey PVL (5%) [19] our device shows a clear improvement while keeping the ability of far-field AM beaming.

In order to better investigate the transmittance efficiency of a Ni tip we performed the simulations where the tip is only partially covered with Ni. In Fig. 3(b) the transmittance of our structure for the cases of $l = 0$ and $l = 1$ are reported as a function of the height (y) of Au (see the inset of Fig.3(b)). As expected the transmittance increase rapidly when the Au coverage grows. The last point corresponds to the situation when only the final 1.5 μm of the tip apex are covered with the nickel. As recently discussed [6, 31], the tip profile we presented was optimized to maximize not only the transmittance, but also the polarization contrast. The latter was defined as $Q = 1 - P_+/P_-$, with P_+ and P_- being the light powers decoupled by the tip with right and left circular polarization state, respectively. Here we want to study this behavior when the Au tip is replaced by a magnetic one. Plots of Q are reported in Fig. 3(c) and (d). As expected, due to higher losses, Q values with the magnetic tips are lower respect to the case of a gold tip. This can be explained by a plasmon depolarization due to the high losses imposing short coherence length. To be specific, the calculated coherence length of a plasmonic wave on a flat magnetic is ~*3* μm, which is roughly a half of the total cone length. As can be seen in Fig. 3(b) Q increases with the PV l order due to the corresponding detachment height decrease. The Q drops again for the PV with $l = 4$ due to a broad spatial divergence of the beam and an additional scattering from Au slab/magnetic tip interface.

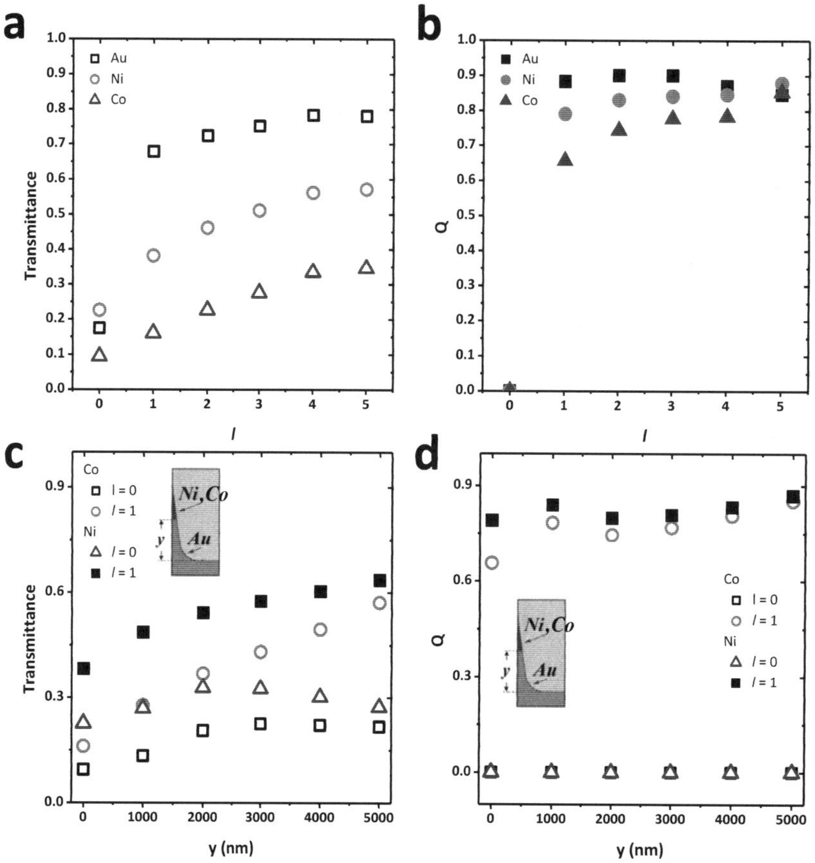

Fig. 3. Comparison of the far-field transmittance (a) and polarization contrast Q (b) obtained from the magnetics tip compared with data obtained with a gold nanostructure [6]; Tip transmittance (c) and Q (d) when the magnetic material covers the upper part of the tip starting from a height denoted by y (see the insets). Computation from $y = 0$ nm (the whole tip is covered by Nickel) to y = 5000 nm (only the upper 1.5 μm of the tip is covered by Ni or Co).

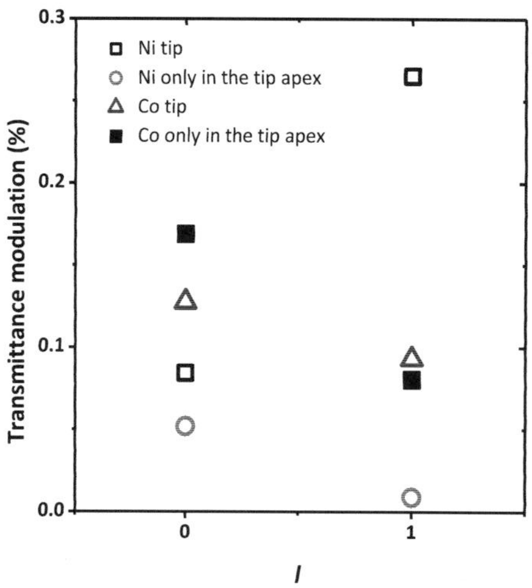

Fig. 4. Magnetic modulation of the total power transmittance for $l = 0$ and 1.

Finally, we study the effect of the application of an external magnetic field. In this case the dielctric tensor of the magnetic material is no more diagonal [30,31]. In Fig. 4 we show the magnetic modulation of the total power transmittance for the cases $l = 0,1$ for both Co and Ni tip, and for the cases where only the final 1.5 μm of the tip apex is covered with the magnetic material. The magnetic modulation is defined as [2*(T(+H)-T(-H))/ (T(+H)+T(-H))]. As it can be seen the modulation is below 1% in all the cases. It is worth noticing that, while in the Ni case covering just the tip apex gives a smaller modulation for both the values of the topological charge considered here, in the Co case the modulation is higher for $l = 0$ when we cover only the tip apex with the magnetic material. Moreover, for $l = 1$, in the Ni case we have almost zero modulation if we cover only the tip apex, while in the case of a pure Ni tip we observe a modulation of almost 0.6%. In the Co case we do not observe a huge change if we have a pure Co tip or a gold tip covered with Co only on the apex. This latter case is interesting because one can use a tip made of pure gold and cover it only in the apex with the magnetic material, thus retaining the good transmittance efficiency (>50%) while maintaining a magnetic modulation functionality. To improve the modulation one can use magnetic materials with a strong magneto-optical activity, such as CoFeTb [31]. The proper design of the system, however, should consider simultaneously optimizing the magnetic and the plasmonic properties of the material. Using these considerations the conical tip structure can be further optimized.

3. Conclusions

In summary, we studied a plasmonic vortex lens structure capable of coupling a circularly polarized light to a plasmonic vortex and efficiently transmitting it to the far-field by means of a smoothed-cone magnetic tip placed at its center. As for the case of Au tip, the large curvature radius at the cone basis was shown to play a crucial role in enabling an adiabatic coupling of the plasmonic vortex propagating on the flat metal surface to the plasmonic modes of the metal tip, whose tapering, in turn, enables an adiabatic match to the propagating waves in free space. The higher losses related to the use of a magnetic material lead to lower transmitted power to the far-field although the beaming is preserved. Changing the configuration by partially covering a Au tip with Ni or Co (on the tapered) can increase significantly the efficiency of the structure. Finally, this particular design shows interesting behaviours when the polarization contrast is evaluated. Finally the magneto-optical modulation effects shown here, once an external magnetic field is applied, suggest intriguing active optical phenomena.

References

[1] Cardano F, and Marrucci L 2015 *Nat. Photonics* **9** 776–778
[2] Tamburini F, Anzolin G, Umbriaco G, Bianchini A, and Barbieri C 2006 *Phys. Rev. Lett.* **97** 163903
[3] Padgett M, and Bowman R 2011 *Nat. Photonics* **5** 343–348
[4] Toyoda K, Miyamoto K, Aoki N, Morita R, and Omatsu T 2012 *Nano Lett.* **12** 3645–3649
[5] Bozinovic N, Yue Y, Ren Y, Tur M, Kristensen P, Huang H, Willner A E, and Ramachandran S 2013 *Science* **340** 1545–1548
[6] Garoli D, Zilio P, Gorodetski Y, Tantussi F, and De Angelis F 2016 *Nano Lett.* **16** 6636-6643
[7] Gorodetski Y, Niv A, Kleiner V, and Hasman E 2008 *Phys. Rev. Lett.* **101** 043903
[8] Gorodetski Y, Shitrit N, Bretner I, Kleiner V, and Hasman E 2009 *Nano Lett.* **9** 3016–3019
[9] Kim H, Park J, Cho S, Lee S, Kang M, and Lee B 2010 *Nano Lett.* **10** 529–536
[10] Bliokh K Y, Rodríguez-Fortuño F J, Nori F, and Zayats A V 2015 *Nat. Photonics* **9** 796–808
[11] Miao J, Wang Y, Guo C, Tian Y, Zhang J, Liu Q, Zhou Z, and Misawa H 2011 *Plasmonics* **7** 377–381
[12] Chen W, Abeysinghe D C, Nelson R L, and Zhan Q 2010 *Nano Lett.* **10** 2075–2079
[13] Cho S-W, Park J, Lee S-Y, Kim H, and Lee B 2012 *Opt. Express* **20** 10083–10094
[14] Zilio P, Mari E, Parisi G, Tamburini F, and Romanato F 2012 *Opt. Lett.* **37** 3234–3234
[15] Zilio P, Parisi G, Garoli D, Carli M, and Romanato F 2014 *Opt. Lett.* **39** 4899–902
[16]Yu H, Zhang H, Wang Y, Han S, Yang H, Xu X, Wang Z, Petrov V, Wang J 2013 *Sci. Rep.* **3** 3191
[17] Genevet P, Lin J, Kats M A, and Capasso F 2012 *Nat. Commun.* **3** 1278
[18] Garoli D, Zilio P, Gorodetski Y, Tantussi F, and De Angelis F 2016 *Sci. Rep.* **6** 29547
[19] Gorodetski Y, Drezet A, Genet C, and Ebbesen T W 2013 *Phys. Rev. Lett.* **110** 203906

[20] Temnov V V 2012 *Nat. Photonics* **6** 728
[21] Armelles G, Cebollada A, García-Martín A, and Ujué González M 2013 *Adv. Opt. Mater.* **1** 10
[22] Maccaferri N, Berger A, Bonetti S, Bonanni V, Kataja M, Hang Qin Q, van Dijken S, Pirzadeh Z, Dmitriev A, Nogués J, Åkerman J, and Vavassori P 2013 *Phys. Rev. Lett.* **111** 167401
[23] Lodewijks K, Maccaferri N, Pakizeh T, Dumas R K, Zubritskaya I, Åkerman J, Vavassori P, and Dmitriev A 2014 *Nano Lett.* **14** 7207–7214
[24] Maccaferri N, Kataja M, Bonanni V, Bonetti S, Pirzadeh Z, Dmitriev A, van Dijken S, Åkerman J, Vavassori P 2014 *Phys. Status Solidi A* **211** 1067-1075
[25] Maccaferri N, Gregorczyk K E, de Oliveira T V A G, Kataja M, van Dijken S, Pirzadeh Z, Dmitriev A, Åkerman J, Knez M, Vavassori P 2015 *Nat. Commun.* **6** 6150
[26] Zubritskaya I, Lodewijks K, Maccaferri N, Mekonnen A, Dumas R K, Åkerman J, Vavassori P, Dmitriev A 2015 *Nano Lett.* **15** 3204–3211
[27] Maccaferri N, Inchausti X, García-Martín A, Cuevas J C, Tripathy D, Adeyeye A O, Vavassori P 2015 *ACS Photonics* **2** 1769-1779
[28] Kataja M, Pourjamal S, Maccaferri N, Vavassori P, Hakala T K, Huttunen M J, Törmä P, van Dijken S 2016 *Opt. Express* **24** 3652-3662
[29] Maccaferri N, Bergamini L, Pancaldi M, Schmidt M, Kataja M, van Dijken S, Zabala N, Aizpurua J, Vavassori P 2016 *Nano Lett.* **16** 2533-2542
[30] Maccaferri N, González-Díaz J B, Bonetti S, Berger A, Kataja M, van Dijken S, Nogués J, Bonanni V, Pirzadeh Z, Dmitriev A, Åkerman J, Vavassori P 2013 *Opt. Express* **21** 9875-9889
[31] Maccaferri N, Gorodetski Y, Toma A, Zilio P, De Angelis F, Garoli D 2017 *Appl. Phys. Lett.* **111** 201104

IOP Conf. Series: Journal of Physics: Conf. Series **961** (2017) 012002 doi:10.1088/1742-6596/961/1/012002

Surface holograms for sensing application

M Zawadzka and I Naydenova

Centre for Industrial and Engineering Optics, School of Physics, Clinical and Optometric Sciences, College of Sciences and Health, Dublin Institute of Technology, Kevin Street, Dublin 8, Ireland

izabela.naydenova@dit.ie

Abstract. Surface gratings with periodicity of 2 μm and amplitude in the range of 175 and 240 nm were fabricated in a plasticized polyvinylchloride doped with a metalloporphyrin (ZnTPP), via a single laser pulse holographic ablation process. The effect of the laser pulse energy on the profiles of the fabricated surface structure was investigated. The sensing capabilities of the fabricated diffractive structures towards amines (triethylamine, diethylamine) and pyridine vapours were then explored; the holographic structures were exposed to the analyte vapours and changes in the intensity of the diffracted light were monitored in real time at 473 nm. It was demonstrated that surface structures, fabricated in a polymer doped with a metalloporphyrin which acts as analyte receptor, have a potential in sensing application.

1. Introduction

Two research problems are being addressed in the herein paper. First concerns fabrication of surface holographic gratings, while the second focuses on exploring the sensing potential of such structures.

Surface holograms can be fabricated using various approaches. Surface relief structures can be formed upon exposure of a photoresist to a patterned light (holographically or with a mask), which process is followed by wet processing needed to develop the surface profile. In mass production, the photoresist structures are copied to create a master which is used in embossing.[1] Surface relief structures can also be created by direct inscription in self-processing photopolymers, but this technique is limited to low spatial frequencies.[2] Post-recording thermal treatment [3] can produce surface profiles at higher spatial frequencies, but this process is time and energy consuming and thus more difficult to implement on a large production scale. Another method of surface hologram fabrication consists of holographic patterning of a material using high power laser which ablates locally material from the surface. This technique is a flexible one-step patterning method that can be applied to a range of materials.[4] The herein paper utilises this technique in patterning of dye doped polymer material. The potential applications of surface structures created in dye doped polymers range from holographic sensors, security devices [5] to distributed feedback lasers (DFB) [6] and DFB lasers based sensors [7].

The herein paper explores the sensing capabilities of the produced structures. The dye used to dope the polymer in which the holographic structure is inscribed belongs to a metalloporphyrin class of dyes which are known to exhibit good binding properties towards various analytes including volatile organic compounds. The sensing capabilities of metalloporphyrins were widely explored in various types of sensors [8], yet to our knowledge they have not been investigated in surface holographic sensors before. The sensing principle of such sensors relies on the change in the refractive index of the sensing material and/or swelling of the grating, which induce changes in optical characteristics of the light diffracted by the surface structure. The operation of the herein sensor is based on detecting changes in diffraction efficiency of the holographic grating, occurring upon its exposure to amines (diethylamine and triethylamine) and pyridine vapours.

2. Experimental details

2.1. Materials

High molecular weight polyvinylchloride (PVC), dioctyl terephthalate (DOTP), $Zn(acac)_2$ and 5,10,15,20-tetraphenylporphyrin (TPP) were obtained from Sigma-Aldrich. Diethylamine (DEA) and pyridine (Pyr) were purchased from Fisher Scientific, while THF, DCM and triethylamine (TEA) were supplied by Roth. All chemicals were used as received. 5,10,15,20-tetraphenylporphyrin zinc(II) (ZnTPP) was synthesized by metallation of TPP with $Zn(acac)_2$ in DMC under reflux using a procedure adapted from literature.[9]

2.2. Fabrication and characterisation of plasticized PVC films doped with ZnTPP

Plasticized PVC films doped with ZnTPP were fabricated by spin coating 100 µl of polymer solution on half of a microscopic glass slide (3000 rpm, 30 s). The polymer solution was prepared by mixing together 75 mg of PVC, 15 mg of DOTP and 6.6 mg of ZnTPP in 3 ml of THF. The thickness of the film was measured with Asylum Research MFP Bio AFM to be 174±20 nm. UV-vis spectra of the films were recorded with Perkin Elmer Lambda 900 UV/VIS/NIR Spectrometer.

2.3. Fabrication of surface gratings via holographic ablation and their characterisation

Surface gratings were fabricated in PVC films doped with ZnTPP using holographic set-up presented in figure 1(a). Ns pulsed Nd:YAG laser (Continuum Surelite) operating at 532 nm was used in this studies. The ablation was carried out with a single laser pulse. The energy of the laser pulse is an average laser pulse energy measured with a pyroelectric detector (Newport, 919E) over 30 s exposure at the laser repetition rate of 10 Hz. The surface profiles of the fabricated structures were investigated with Nanosurf AFM. To determine diffraction efficiency of the 1st order (figure 1(b)), the structures were illuminated with a diode laser operating at 473 nm and the intensity of the diffracted 1st order beam I_1 was measured with a photodetector (Newport, 818-SL/DB); diffraction efficiency η was calculated from $\eta = I_1/I_0$ where I_0 is the intensity of the incident beam.

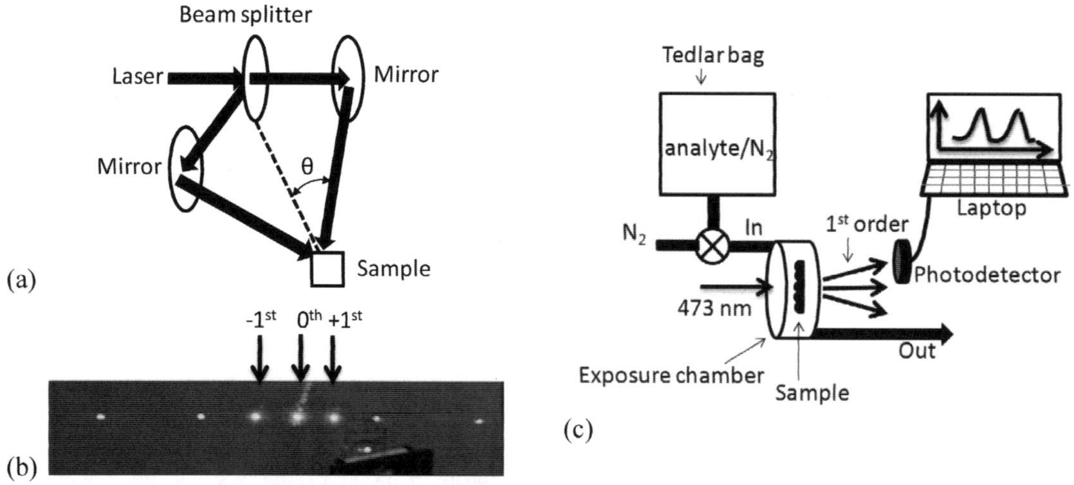

Figure 1. (a) Set-up used to holographically pattern the material (θ =7.6°) (b) photography of the light pattern obtained upon illumination of the fabricated structure with 473 nm laser with indicated ±1st and 0th orders (c) set-up used to investigate response of the holographic structures to analyte vapours.

2.4. Experimental set-up used to examine the response of the holographic structure to analytes vapours

The scheme of the set-up used to investigate the sensing response of the holographic structures to TEA, DEA and Pyr vapours is illustrated in figure 1(c). The sample was placed in a 100 ml chamber connected to a gas inlet switchable between pure N_2 and mixture of the analyte vapours and N_2. The outlet of the chamber was connected to the pump. The mixtures of the analytes and N_2 at the analyte concentration of 10 000 ppm were prepared by injecting the appropriate volume of the liquid analyte into a 10 L Tedlar bag, which was then followed by filling the bag with N_2.

The general procedure for the investigation of the response of the holographic structure to analyte was as follow. First, the sample was exposed to N_2 flow (1 L/min); once the signal stabilized the inlet gas was switched to the analyte mixture; then after the response saturated, N_2 flow was turned on again in order to study the recovery of the initial signal and check the reversibility of the sensor's response. The signal, being the intensity of the diffracted 1st order at probing wavelength of 473 nm, was monitored in real time with a photodetector connected to the computer interfaced with software for data acquisition.

3. Results and discussion

3.1. Surface holographic gratings fabricated through single pulse ablation

Ablation, in general, depends simultaneously on the properties of the laser and these of the material. The decomposition of the material can go through various pathways depending on the nature of the interaction of the light with the material. For ns pulsed lasers, ablation is often driven by thermal processes and it is wavelength dependent.[10] In our studies, the ablating laser wavelength of 532 nm coincides with edge of the absorption peak corresponding to the Q-band of ZnTPP (figure 2(a)). The absorbance of the polymer film at 532 nm was found to be 0.0177.

Employment of ablation in holographic patterning combines the advantages of a one-step fabrication process with the flexibility of the experimental set-up which can be easily adjusted to obtain structures with the desired periodicity. Since periodicity of the grating often limits the achievable thickness of the grating d, and diffraction efficiency η depends on d as per equation (1) (thin grating regime), the periodicity of the grating must be adjusted to obtain structures of the desired d and thus η.

$$\eta = Ji^2(\Phi) \tag{1}$$

$$\Phi = 2\pi\Delta n d/\lambda \tag{2}$$

where Ji is the Bessel function of the first kind, Φ is phase, Δn is refractive index modulation of the grating and $\Delta n = n_g - n_m$ with n_g being the refractive index of the grating material and n_m - refractive index of the medium surrounding the grating, and λ is wavelength of the interrogating laser.

For the periodicity of 1 μm, the achievable d was on the order of 100 nm. Increasing periodicity to 2 μm allowed formation of structured with d = 240 nm. The maximal d was obtained upon optimisation of the energy of the ablating laser pulse. The results of such optimisation for structures with the periodicity of 2 μm (being of main interest in the herein studies), are illustrated in figure 2(b). It can be seen from figure 2(b) that d and η follow similar trend, in accordance with equation (1). It should be mentioned that formation of the surface structure became already visible at the laser pulse energy ~ 10 mJ, yet structures with highest η = 28% were obtained with laser pulse energy reaching 133 mJ. η = 28% is close to the theoretical maximum (33.5%) for thin holographic surface grating and was obtained for the structures with d = 240 nm. An example of data on surface studies obtained with AFM is illustrated in figure 2(c).

Apart from studies discussed above, optimisation of the structures was also carried out through adjustment of the composition of the material. The starting composition was based on the formulation often used in fabrication of sensors, with 1:2 weight ratio between PVC and plasticizer, respectively.[11] Such highly plasticized composition was reported to result in quick response time of a sensor. Although, it was possible to fabricate holographic structures in materials with such high

IOP Conf. Series: Journal of Physics: Conf. Series **961** (2017) 012002 doi:10.1088/1742-6596/961/1/012002

plasticizer content, the structures were unstable over the time (η was noted to decrease over the time). Using composition with 5:1 (PVC:DOTP) ratio allowed fabrication of stable structures which thought might suffer from slower response and recovery times.

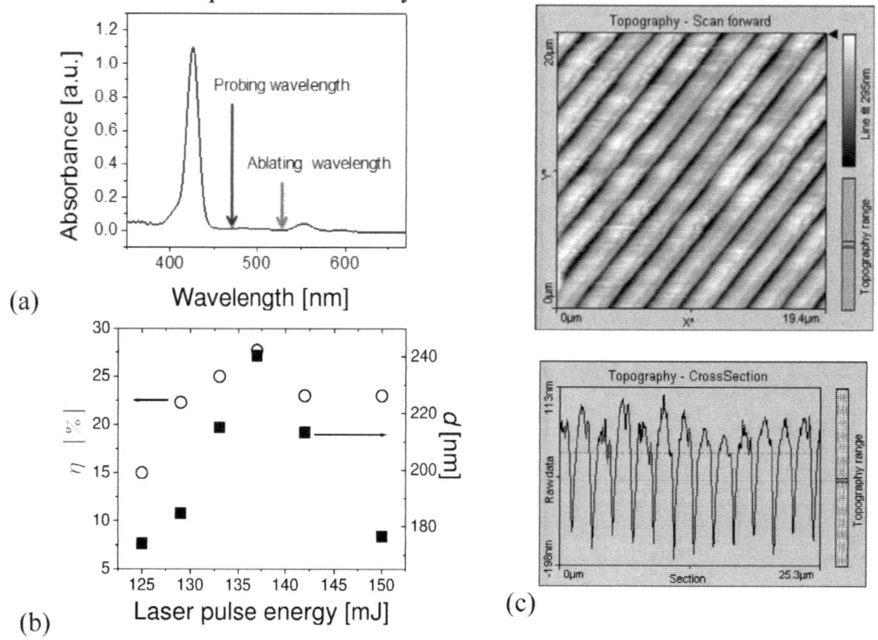

Figure 2. (a) UV-vis spectrum of the polymer film (b) diffraction efficiency η and thickness of the grating d plotted versus energy of the laser pulse used to fabricate surface holograms through ablation (c) AFM data showing surface map (upper) and cross-section of the grating (lower).

As illustrated above, holographic patterning based on ablation offers quick fabrication of efficient surface structures. The limitation of the technique is the size of the structures, which is determined by a need for high power laser irradiation necessary for the ablation process. The diameter of the structures fabricated in the herein studies was ~ 6 mm.

3.2. Sensing capabilities of the produced structures towards TEA, DEA and Pyr vapours

Metalloporphyrins are good receptors of many volatile organic compounds, which capability was broadly employed in colorimetric sensors.[12] The question of the herein studies was whether their binding capabilities can be used in surface holographic structures to produce effective sensors. The working principle of such sensors is based on the change in the refractive index which arises in the material in presence of the analyte and/or swelling of the structure (equation (1)).

The changes in η occurring upon exposure of the structures to analytes was monitored at λ = 473 nm. This λ is in the spectral window of both ZnTPP and ZnTPP coordinated by the analytes.[13] (Alternatively, interrogation could be carried out at $\lambda > 650$ nm, but for the same grating thickness a lower diffraction efficiencies will be detected, as seen from equation (2)).

Table 1 summarises the responses of the holographic structures to the three analytes. No changes in η where observed upon exposure to DEA, while an increase by 1 and 3% in η was noted for TEA and Pyr, respectively. The response to vapours was immediate, yet saturation of the signal occurred in 10 min for TEA and 5 min for Pyr. No full recovery of the baseline could be achieved upon subsequent exposure of the holographic structure to pure N_2 in the following step (up to 30 min) for TEA or Pyr; it could be seen that the colour of the samples outside holographic structures differed from that of the fresh sample. This indicated that the analytes were not affectively removed from the layer, which problem could potentially be solved by incorporation of a heater to the sensor's architecture. Subsequent exposure of the grating to the analyte vapours resulted again in an increase of the signal (figure 3).

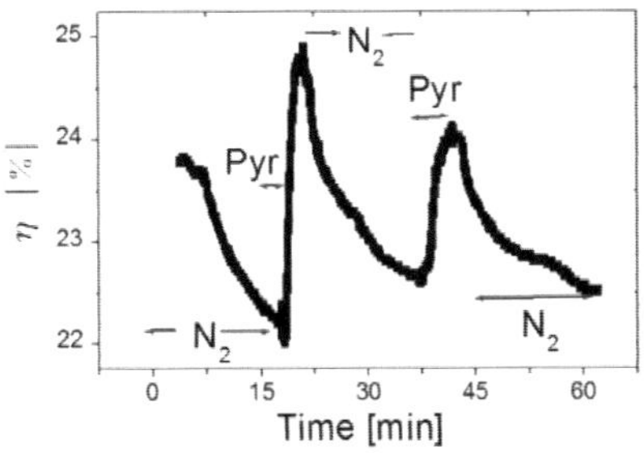

Figure 3. Changes in η monitored over the time upon exposure of the holographic structure to the flow of N_2 followed by flow of Pyr/N_2 mixture (two cycles of exposure).

Table 1. Summary of the responses of the surface structures to analytes.

Analyte	$\Delta\eta$ [%]	Time of max $\Delta\eta$ [min]
Pyr	3	5
TEA	1	10
DEA	0	-

It should be mentioned that structures with η = 24% (and not with max η) were used in sensing studies; this should provide larger dynamic range for sensor operation, which will be illustrated in the following part (figure 4).

3.3. Simplified theoretical analysis of the sensing response

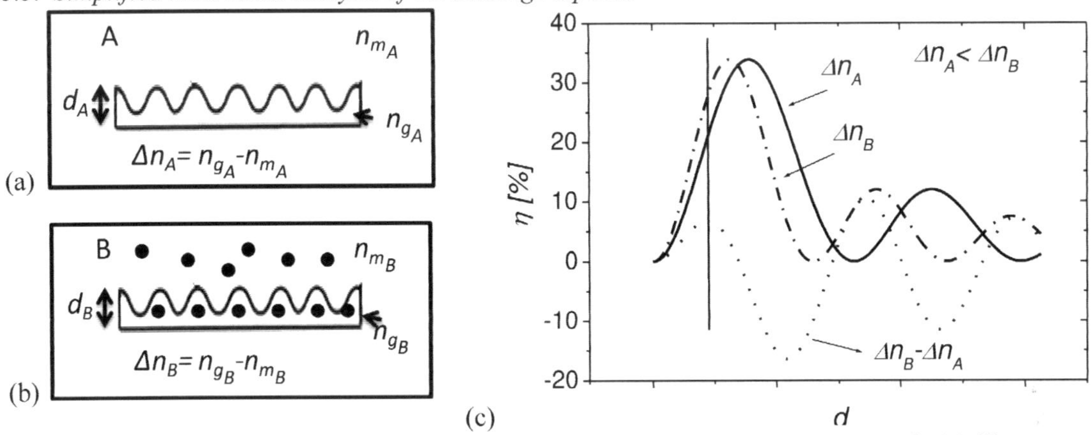

Figure 4. (a) and (b) illustrate grating before and after exposure to analyte. Graph (c) illustrates η plotted versus *d* (equation (1)) (similar graph could be obtained for $\eta = f(\Delta n)$). It is assumed that changes in *d* are negligible in comparison to changes in Δn. For a given thickness *d* indicated with a vertical line in the plot, η will increase if Δn increases provided that $\Delta n_B - \Delta n_A > 0$. Optimal operation regime is where $\Delta n_B - \Delta n_A$ reaches max or min. (The figure is just for illustration purpose and it is highly simplified; other interpretations including variation in surface amplitude are also possible.)

To fully understand the sensing response and the observed trend for the three analytes, further studies on refractive index of the whole system under investigation are needed and the part below only speculates about possible processes responsible for the observed responses. The response of the herein grating is governed by equation (1). As can be seen from equation (2) the change in the diffraction efficiency η can be caused by either a change in the thickness of the grating *d* or a change in the refractive index modulation Δn. Exposure of the polymer films to analytes can induce swelling thus the thickness of the grating *d* can potentially increase. As far as changes in the refractive index modulation Δn arising due to exposure of the film to the analyte are concerned, there are two contributes n_g and n_m. Introduction of analytes vapours to the medium surrounding the grating (N_2) will result in an increase of n_m. Changes in n_g are more difficult to predict; introduction of the analyte into polymer matrix should lead to increase of n_g while material swelling will have the opposite effect. Taking into account experimental data from table 1 and comparing them against the refractive index of the analyte, it can be seen that trend in η observed experimentally for the three analytes $\Delta\eta$(Pyr) > $\Delta\eta$(TEA) > $\Delta\eta$(DEA) = 0 is correlated to refractive index of the analytes (*n*(Pyr) =1.509 > *n*(TEA)

=1.401 > n(DEA) =1.385). Based on this, it can be speculated that this is an increase in Δn which governs the response with the main contribution coming from n_g. Figure 4(c) depicts the predicted operation regime of the sensor. Further experimental and modelling studies are yet needed to confirm and to fully understand the response. Measurements of the refractive index of the unpattern layer under exposure to the analyte will help understanding the observed phenomena.

4. Summary and future perspectives

It was demonstrated that holographic ablation is a quick and flexible fabrication technique of surface holographic gratings in functionalized polymer materials. The sensing capabilities of the fabricated structures to volatile organic compounds such as amines and pyridine were explored and presented. Although further studies and improvements on the characteristics of the holographic structures' responses to analytes are needed to obtain competitive sensors, the results reveal still unexplored application potential of surface holograms in sensing. Other potential and important applications of surface holograms fabricated in functionalized materials could include stimuli responsive security devices and actuators.

5. References

[1] Colburn W S 1997 *J. Imaging Sci. Technol.* **41** 443
[2] Naydenova I, Mihaylova E, Martin S and Toal V 2005 *Opt. Express* **13** 4878
[3] Trainer K, Wearen K, Nazarova D, Naydenova I and Toal V 2010 *J. Opt.* **12** 124012
[4] Yetisen A K, Qasim M M, Nosheen S, Wilkinson T D and Lowe C R 2014 *J. Mater. Chem. C* **2** 3569; Alqurashi T, Montelongo Y, Penchev P, Yetisen A K, Dimov S and Butt H 2017 *Nanoscale* **9** 13808; Kuang Z, Perrie W, Liu D, Fitzsimons P, Edwardson S P, Fearon E, Dearden G and Watkins K G 2012 *Appl. Surf. Sci.* **258** 7601
[5] Yetisen A K, Naydenova I, da Cruz Vasconcellos F, Blyth J and Lowe C R 2014 *Chem. Rev.* **114** 10654
[6] Stroisch M, Woggon T, Lemmer U, Bastian G, Violakis G and Pissadakis S 2007 *Opt. Express* **15** 3968
[7] Tan Y F, Ge C, Chu A, Lu M, Goldshlag W, Huang C S, Pokhriyal A, George S and Cunningham B T 2012 *IEEE Sens. J.* **12** 1174
[8] Chou J H, Kosal M E, Nalwa H S, Rakow N A and Suslick K 2000 *The porphyrin handbook. vol 6, Applications : past, present and future* ed Kadish K M, Smith K M and Guilard (San Diego, Calif. ; London: Academic) p 114
[9] Zawadzka M, Wang J, Blau W J and Senge M O 2012 *J. Phys. Chem. A* **117** 15
[10] Stafe M, Marcu A and Puscas N N 2013 *Pulsed Laser Ablation of Solids: Basics, Theory and Applications* (Springer Berlin Heidelberg) p 40
[11] Vlascici D, Popa I, Chiriac V A, Fagadar-Cosma G, Popovici H and Fagadar-Cosma E 2013 *Chem. Cent. J.* **7** 111
[12] Askim J R, Mahmoudi M and Suslick K S 2013 *Chem. Soc. Rev.* **42** 8649-82
[13] Rukin P S, Kashchenko P A, Malyavskaya A Y, Bagaturyants A A and Alfimov M V 2014 *Nanotechnol. Russia* **9** 136-44

Acknowledgements

This research was funded by Irish Research Council Government of Ireland Postdoctoral Fellowship Scheme grant no. GOIPD/2015/743. FOCAS institute at Dublin Institute of Technology is acknowledged for providing laboratory resources. The authors would like to thank Prof. John Cassidy and Dr Jack Treacy for their advice on the development of the experimental set-up for the characterization of sensing capabilities of the holographic structures, and Dr Luke O'Neil for measuring the thickness of polymer films.

Effect of temperature on the spectral properties of InP/ZnS nanocrystals

S S Savchenko, A S Vokhmintsev and I A Weinstein

Ural Federal University, NANOTECH Centre, Mira street, 19, Yekaterinburg, Russia, 620002

E-mail: i.a.weinstein@urfu.ru

Abstract. Optical absorption (OA) and photoluminescence (PL) spectra of InP/ZnS core/shell nanocrystals with 2.3 nm average size were investigated in the temperature range of 6.5–296 K. Using second derivative spectrophotometry technique energies of the OA transitions at 296 K in quantum dot (QD) solutions and films are evaluated to be E_1 = 2.37, E_2 = 4.10 and E_3 = 4.68 eV. Temperature shifts of the E_1 and E_2 levels are found to result from interaction with effective phonons of 59 and 37 meV energies, respectively. Herewith the 370 meV half-width of the first exciton absorption peak remains constant due to the dominance of inhomogeneous broadening effects caused by QD parameters distribution. Measured PL spectra have a complex structure and can be described in 6.5–296 K range by two independent Gaussian components associated with exciton and defect-related states. In addition, Stokes shift of 320 meV is observed to decrease at $T > 200$ K. PL thermal quenching analysis in frame of Mott mechanism points to presence of non-radiative relaxation channel with an activation energy of 74 meV.

1. Introduction

Great interest has been focused on $A^{III}B^{V}$ colloidal quantum dots (QDs) since they possess tunable size-dependent optical properties and low toxicity. Among this group, InP QDs coated with zinc sulfide shell exhibit characteristics good enough to substitute environmentally restricted Cd-based counterparts [1,2]. It makes InP/ZnS QDs promising both in optoelectronic and biocompatible applications on an industrial scale [3,4]. At the same time, optical properties of InP/ZnS nanocrystals remains insufficiently studied both from the standpoint of experimental analysis and consistent theoretical justifications. In this regard, temperature-dependent optical absorption (OA) and photoluminescence (PL) measurements could help to obtain valuable information on emission and absorption exciton states. To date, there are several studies reporting the temperature analysis of InP/ZnS PL spectra and investigating their band-edge exciton fine structure [5–7]. Previously, we have studied OA temperature dependence in QDs of 2.1 nm average diameter [8,9]. This work aims to investigate temperature behavior of OA and PL spectra of larger InP/ZnS QDs of 2.3 nm size in the range of 6.5–296 K.

2. Samples and Techniques

InP/ZnS colloidal QDs under study marked as DS-72b were synthesized by Institute of Applied Acoustics (Dubna, Russia). They have three-layered structure: the core made of InP, the shell made of ZnS, and the coating of modified polyacrylic acid. Semiconductor nanocrystals were dispersed in water with initial concentration of 48.4 g/l. An average core diameter estimated from the first exciton energy [10] is 2.3 nm. Quantum yield of the nanocrystals was determined using relative procedure with Rhodamine 6G as a reference [11] and amounted to 10 %.

Optical absorption measurements of samples were carried out in a range from 190 to 900 nm by means of a Shimadzu UV-2450 spectrophotometer. Solutions of different concentrations were investigated at room temperature (RT) in a quartz cuvette with 1 cm optical path. To establish the temperature dependence of the nanocrystals OA spectra, we used InP/ZnS film formed by physical deposition of the initial colloidal solution at room temperature on a quartz substrate with a thickness of 1 mm. The substrate transmission is 94 % in the 300-900 nm range smoothly decreasing to 85 % at 190 nm. Absorption spectra at T = 6.5–296 K were measured using a combined installation based on the Shimadzu UV-2450 spectrophotometer and a Janis closed cycle refrigerator system, model CCS-100/204N, equipped with a DT-670B-CU temperature sensor and Model 335 controller. A HiCube 80 Eco turbo pumping station maintained vacuum inside the cryostat ($7\cdot10^{-5}$ mbar). The spectra were registered at the following temperature values: 6.5 K, over the range from 10 to 100 K in 10 K, and in the range of 100–296 K with a temperature step of 20 K. A slit width and sampling interval were set to the 2 and 0.5 nm, respectively.

Photoluminescence spectra of the QDs were analyzed for InP/ZnS film in the range 6.5–296 K. For this purpose, the described refrigerator system was coupled with Andor Technology Shamrock SR-303i-B spectrograph and Newton^{EM} DU-970P-BV-602 CCD-camera. LED with emission maximum at 372 nm was used as an excitation source.

3. Results

Figure 1 shows optical absorption spectra for solutions of InP/ZnS QDs with 40 and 1.25 g/l concentrations as well as for films at 6.5 and 296 K. For clarity, the data is presented in semilogarithmic coordinates. Three features, indicated by arrows in figure 1, can be distinguished on spectra. E_1 and E_2 shoulders in the ranges of 490–570 nm and 290–330 nm, respectively, are observed for all samples in the solution and film form. The E_3 feature in the range of 250–280 nm appears only for solutions with a low concentration. As for the temperature behavior of the spectra, the positions of the optical transitions are shifted toward higher energies when T is lowered. Simultaneously, the optical density increases within the investigated wavelength range.

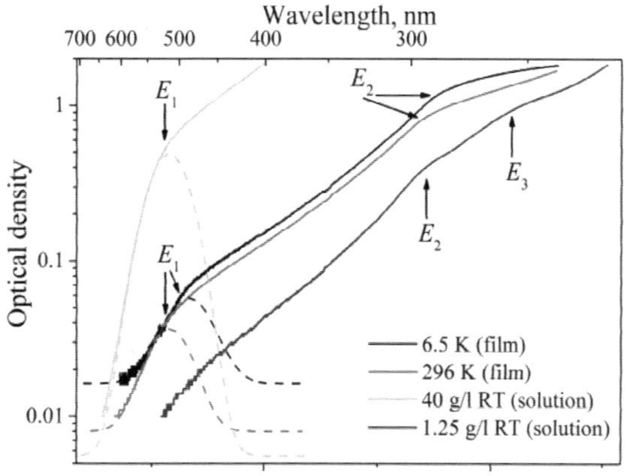

Figure 1. OA spectra of the InP/ZnS at various concentrations and temperatures. Dashed lines represent Gaussian approximation of the first exciton band.

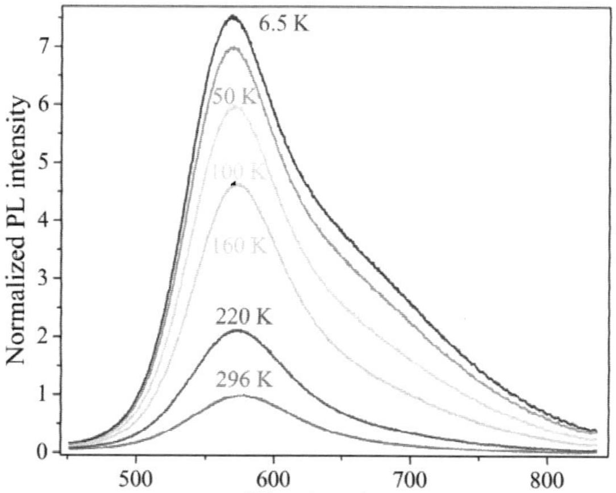

Figure 2. PL spectra of the InP/ZnS at various temperatures.

Figure 2 shows the photoluminescence spectra of the InP/ZnS film registered at different temperatures. The PL intensity is normalized to the maximum value at room temperature. It is seen that the luminescence spectrum at RT is an asymmetric band with a maximum at a wavelength of λ_{max} = 578 nm. Upon temperature decreasing, the peak moves to shorter wavelengths and λ_{max} = 569 nm at 6.5 K. The intensity at the maximum increases by 8 times when temperature drops from 296 to 6.5 K. Note that below 100 K one can clearly observe a spectral shoulder in the range of 630–740 nm.

4. Discussion

4.1. Optical absorption and exciton-phonon interaction

The OA features, indicated by arrows in figure 1, reflect the transitions between the quantized exciton energy levels of nanocrystals. They manifest themselves on a monotonically increasing background associated with the contribution of high-energy states, scattering and other processes. Determination of such hidden bands was carried out using the second derivative of the absorption spectra [9]. Figure 3 illustrates this procedure for the calculation of E_1 and E_2 energies for the case of the InP/ZnS film absorption spectra at 80 and 220 K. Table 1 shows the obtained values. It should be noted that the optical transition energies in solutions do not depend on concentration and coincide with the estimates for the film at room temperature within ±0.02 eV experimental error. The absence of the E_3 = 4.68 eV transition in the film spectra may be associated with the cooperative phenomena in QD arrays, since change in the distance between them during deposition leads to a modification of their energy structure [12].

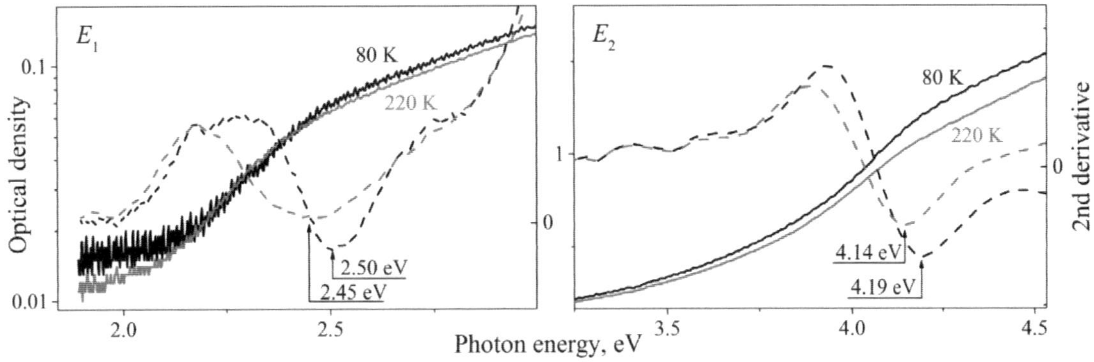

Figure 3. OA spectra (solid lines) and their second derivatives (dash lines) for InP/ZnS film.

As in our work [9], for the quantitative analysis of the experimental dependences E_1 (T) and E_2 (T), we used Fan expression for the temperature behavior of the energy gap E_g in crystals [13]:

$$E_g(T) = E_g(0) - A_F\left[\exp\left(\frac{\hbar\omega}{kT}\right) - 1\right]^{-1}, \tag{1}$$

as well as a semi-empirical linear model:

$$E_g(T) = E_g(0) - \beta T. \tag{2}$$

Here $E_g(0)$ is the energy of the optical transition at 0 K, eV; A_F is the Fan parameter that depends on microscopic properties of material [13,14], eV; $\hbar\omega$ is the effective phonon energy; k is the Boltzmann constant, eV/K; β is the energy gap temperature coefficient, eV/K. As previously demonstrated in [13], in the limit of high temperatures ($kT \gg \hbar\omega$), the relation (1) reduces to (2), with the temperature coefficient being written as:

$$\beta_\infty = A_F \frac{k}{\hbar\omega}. \tag{3}$$

Figure 4 shows approximations of the experimental $E_1(T)$ and $E_2(T)$ dependencies using the Fan relation (1) (blue solid lines) and the linear model (2) (blue dash lines). The obtained parameters are given in table 1. The displacement of the E_1 and E_2 energy levels is found to be due to the interaction with different effective phonon modes. The Huang-Rhys factor S proportional to the strength of the electron-phonon coupling [5,6] and related to A_F as $S = A_F/2\hbar\omega$ [9], is 2.4 times larger for the E_1 band. Figure 4 also represents the dE_g/dT data (red dash lines), which characterizes the change in the temperature coefficient β. The dash-dot lines correspond to the β_∞ level calculated according to expression (3).

For the first exciton absorption peak we also estimated the full width at half maximum H. For this purpose, the low energy region of spectra was approximated by a Gaussian band centered at E_1, see the dashed lines in figure 1. The calculated values showed that the H does not change in the investigated temperature range and amounts to 370±30 meV. Such a result indicates the dominance of the inhomogeneous broadening caused by the distribution of studied nanocrystals in the ensemble in size, shape, imperfections, etc. [9,10,15].

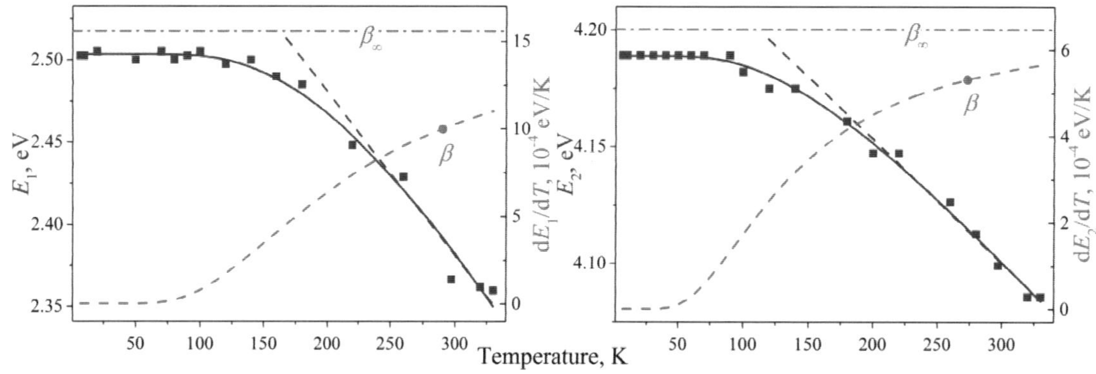

Figure 4. Temperature dependencies for E_1 and E_2. Blue square symbols are our experimental estimates.

Table 1. OA parameters of InP/ZnS QDs.

Optical transition	E, eV (at 296 K)	H, meV (at 296 K)	E, eV (at 6.5 K)	H, meV (at 6.5 K)	A_F, eV	$\hbar\omega$, meV	S	β, 10^{-4} eV/K	β_∞, 10^{-4} eV/K
E_1	2.37	365	2.49	359	1.06	59	9.03	9.97	15.56
E_2	4.10	–	4.19	–	0.27	37	3.76	5.31	6.47

4.2. Exciton and defect-related photoluminescence

All measured PL spectra of InP/ZnS in the investigated temperature range were deconvoluted with a high accuracy (Adj. R-square > 0.999) by two components of a Gaussian shape. Figure 5 shows the approximation examples of experimental data at 6.5 and 296 K. In [16] luminescence spectra of InP/ZnS nanocrystals with a 2 eV emission maximum at 300 K were successfully approximated by two components of a Lorentz shape. The E_{exc} component can be associated with exciton states and the E_d band related to defect states of the QD crystal structure. The exciton energy demonstrates typical blueshift under cooling as presented in figure 6. However, it cannot be fitted with expression (1) owing to the feature in a range of 200–240 K. The defect-related peak is redshifted having temperature-independent half-width of H = 600±20 meV. These facts indicates a possible more complex structure of the PL band and the need to identify origin of the components for an adequate description of the observed temperature behavior.

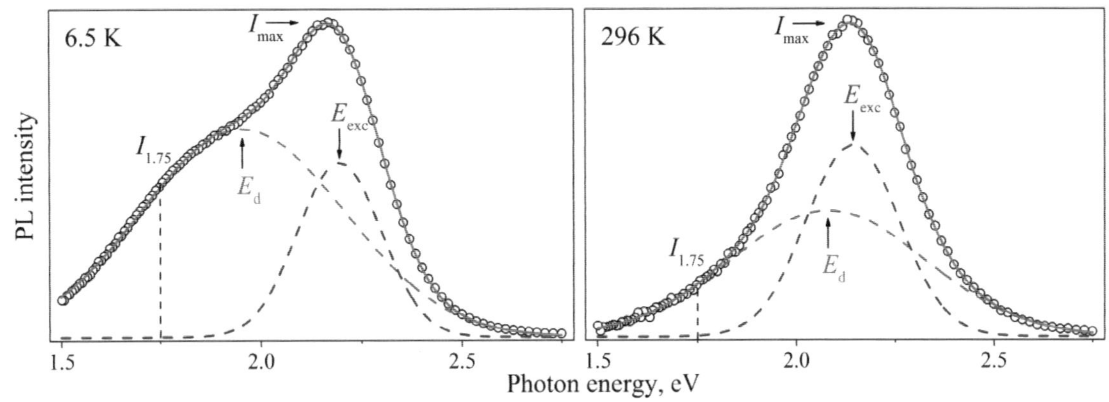

Figure 5. Deconvolution of PL spectra at 6.5 and 296 K. Symbols represent experimental data, solid red lines are fitted curves, dash lines are exciton E_{exc} and defect-related E_d components.

Figure 6 inset shows the temperature dependence of the Stokes shift calculated as $E_1 - E_{exc}$. It is noticed to be almost independent of temperature in the 6.5–200 K range, and then its value begins to decrease. This behavior may be due to the fine structure of exciton emission. The quantum confinement effect enhances the spin-exchange interaction and leads to an increase in the splitting between bright and dark exciton states. The contribution of the dark state with a lower energy to the PL band is maximal at low temperatures. Population of the bright exciton level grows with temperature eventually resulting in a decrease of the Stokes shift from 320 to 220 meV at $T > 200$ K [17].

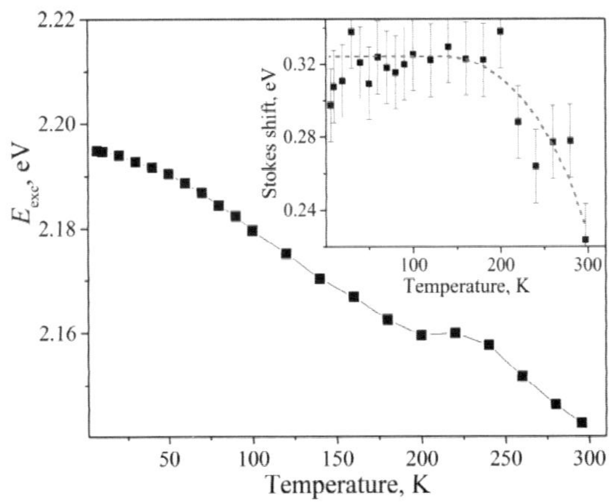

Figure 6. Temperature dependence of the E_{exc} and Stokes shift (inset). Dash curve is guide for the eye.

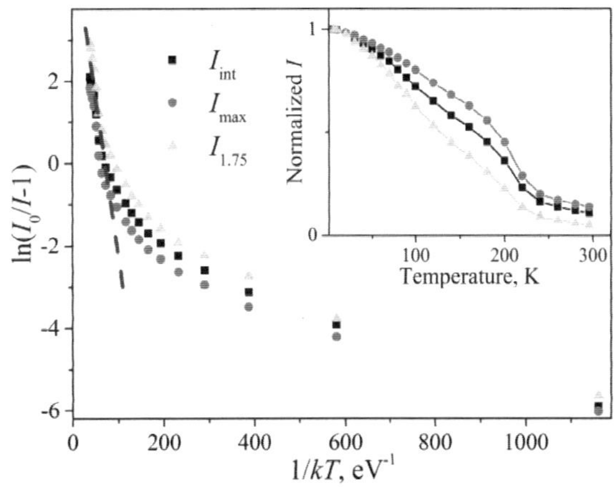

Figure 7. PL temperature quenching. Dash line shows linear approximation.

4.3. PL thermal quenching

To analyze the processes of PL thermal quenching, three parameters were taken: an integrated intensity (I_{int}), luminescence intensities at spectrum maximum (I_{max}) and at 1.75 eV ($I_{1.75}$) which corresponded to the E_{exc} and E_d components (see figure 5). Figure 7 inset shows the PL quenching curves for these quantities normalized at 6.5 K. They exhibit different character of intensity drops with T. Nevertheless, in Arrhenius coordinates it can be seen that all dependencies have a common linear region (dash line), which slope determines the activation energy E_a of PL quenching in frame of Mott relation [18,19]. Thus, there is a common non-radiative relaxation channel above 150 K with E_a =

74±6 meV. Quenching in the low-temperature region cannot be described with a linear function, which points a more complex mechanism.

5. Conclusion

The temperature dependencies of the optical absorption and photoluminescence spectra of the InP/ZnS core/shell QDs with an average diameter of 2.3 nm were studied in a range of 6.5–296 K. Three optical transitions with energies of E_1 = 2.37, E_2 = 4.10 eV and E_3 = 4.68 eV are shown to appear in the spectra of QD solutions and films at room temperature (RT). Analysis of the E_1 and E_2 temperature shifts in the framework of the Fan formalism showed that they are caused by interaction with effective phonons of 59 and 37 meV energies, respectively. Herewith, the half-width of the first exciton absorption peak equals to 370±30 meV and does not change due to the dominance of inhomogeneous broadening resulting from QD distributions in size, shape, imperfections, etc.

Photoluminescence is characterized by broad emission band with 2.14 eV maximum at 296 K. All spectra measured in the 6.5–296 K range are deconvoluted into two components of a Gaussian shape with 2.14 and 2.08 eV maxima at RT. They are assumed to originate from exciton and defect-related states, correspondingly. A PL quenching analysis in frame of Mott relation indicates that there is a common non-radiative relaxation channel with an activation energy of 74 meV. In addition, a decrease in the Stokes shift is observed from 320 to 220 meV at T > 200 K implying large energy splitting between dark and bright exciton states.

Acknowledgments

The work was supported by Act 211 Government of the Russian Federation, contract № 02.A03.21.0006.

References

[1] Brichkin S B 2015 *Colloid J.* **77** 393–403
[2] Rempel S V., Kozhevnikova N S, Aleksandrova N N and Rempel A A 2011 *Inorg. Mater.* **47** 223–6
[3] Xie R, Battaglia D and Peng X 2007 *J. Am. Chem. Soc.* **129** 15432–3
[4] Savchenko S S, Vokhmintsev A S and Weinstein I A 2016 *AIP Conf. Proc.* **1717** 40028
[5] Narayanaswamy A, Feiner L F and van der Zaag P J 2008 *J. Phys. Chem.* C **112** 6775–80
[6] Narayanaswamy A, Feiner L F, Meijerink A and Van Der Zaag P J 2009 *ACS Nano* **3** 2539–46
[7] Biadala L, Siebers B, Beyasit Y, Tessier M D, Dupont D, Hens Z, Yakovlev D R and Bayer M 2016 *ACS Nano* **10** 3356–64
[8] Savchenko S S, Vokhmintsev A S and Weinstein I A 2017 *Tech. Phys. Lett.* **43** 297–300
[9] Savchenko S S, Vokhmintsev A S and Weinstein I A 2017 *Opt. Mater. Express* **7** 354–9
[10] Gaponenko S V 1998 *Optical Properties of Semiconductor Nanocrystals* (Cambridge: Cambridge University Press)
[11] Savchenko S S, Vokhmintsev A S and Weinstein I A 2016 *J. Phys. Conf. Ser.* **741** 012151
[12] Mićić O I, Jones K M, Cahill A and Nozik A J 1998 *J. Phys. Chem.* B **102** 9791–6
[13] Vainshtein I A, Zatsepin A F and Kortov V S 1999 *Phys. Solid State* **41** 905–8
[14] Fan H Y 1951 *Phys. Rev.* **82** 900–5
[15] Salvador M R, Graham M W and Scholes G D 2006 *J. Chem. Phys.* **125** 184709
[16] Pham T T, Chi Tran T K and Nguyen Q L 2011 *Adv. Nat. Sci. Nanosci. Nanotechnol.* **2** 25001
[17] Watanabe T, Takahashi K, Shimura K and Kim D 2017 *Phys. Rev.* B **96** 35305
[18] Weinstein I A, Vokhmintsev A S and Kortov V S 2006 *Tech. Phys. Lett.* **32** 58–60
[19] Curie D 1963 *Luminescence in Crystals* (London: Methuen & Co Ltd / New York: John Wiley & Sons Inc)

IOP Conf. Series: Journal of Physics: Conf. Series **961** (2017) 012004 doi:10.1088/1742-6596/961/1/012004

Analytical Investigation of In-plane Focusing Surface Plasmon Modes by a Dielectric Lens

Fahimeh Armin, Mir Mojtaba Mirsalehi

Department of Electrical Engineering, Faculty of Engineering, Ferdowsi University of Mashhad, Mashhad 9177948974, Iran

E-mail: mirsalehi@um.ac.ir

Abstract. A dielectric lens placed on a metal-dielectric interface can be used for in-plane focusing of surface plasmon (SP) modes. The propagation behind the lens is usually studied using scalar diffraction theory which is not accurate in nanometer regime and cannot explain the propagation and energy flow of SP modes. Here, we have used vectorial derivation of Huygens-Fresnel principal to calculate the diffracted fields behind the lens in order to study the energy flow of SP modes in a focusing system using the concept of higher order modes.

1. Introduction

Local oscillations of conduction electrons at the interface of a metal and a dielectric will form surface plasmons (SPs), an explanation for what Wood observed in 1902 but was not understood till the publication of Ritchi in 1957 [1]. Since then, plasmonic surface waves have found their ways through various fields, from communications to biosensing [2, 3]. In each field, many interesting structures have been studied and proposed in papers.

In some cases, the use of SPs has made it possible to redesign classical optical components and devices in nanoscale. Zia and Brongersma have simulated the Young double slit experiment by plasmon waves [4] while Hohenau *et al.* have experimentally studied the transmission of SPs through prisms and lenses made of SiO_2 [5]. Shi *et al.* studied dielectric components made of Al_2O_3 to manipulate in-plane propagation of surface plasmon modes resulted from coupling of light by a single slit on a silver-air interface [6]. Some papers like the work presented by Bezuse *et al.* studied the transmission of surface plasmons from diffractive optical components in more details by calculating the phase modulation of these modes using rigorous coupled wave analysis [7].

The aforementioned papers show that it is possible to use dielectric components for guiding and shaping SPs like other electromagnetic waves. However, there are more assumptions to be considered in this case, like the propagation medium which is not loss free. Moreover, SPs are p-polarized but inhomogeneous plane waves [8] that means the real and imaginary parts of wavevector, $\boldsymbol{k}$, are not parallel to each other. While analyzing the problem of diffraction from a dielectric lens or other optical components, many papers consider the excitation of scattering or radiation modes behind the dielectric components in which a part of the energy is decoupled from the surface plasmon mode. The concept of excitation of other mode was also studied in other inhomogeneities. For example, Outlan *et al.* used Gaussian quadrature method to solve the

IOP Conf. Series: Journal of Physics: Conf. Series **961** (2017) 012004 doi:10.1088/1742-6596/961/1/012004

mode matching integrals in an open waveguide structure to obtain the transmission, reflection, and also scattering coefficients at the boundary [9].

Kim *et al.* have addressed the issue of losses of the so called scattering modes by adding an air layer between the dielectric lens and metal interface [10]. Bezus *et al.* used the same solution but with a dielectric layer in which optimizing the height of the layer could significantly decrease the losses.

In this paper, we have studied the transmission and diffraction of surface plasmon modes by a dielectric lens using a vectorial Huygens-Fresnel principal to calculate the diffracted fields and then compared them to higher order plasmonic modes. The rest of the paper is organized as follows. Section 2 is devoted to describe the vectorial Huygens-Fresnel method used for calculation of diffracted fields. The argument of higher order plasmonic modes behind a dielectric lens is presented in Section 3 along with the study of losses imposed by these modes. Finally, Section 4 is dedicated to conclusions.

2. Calculations of diffracted fields

For a surface plasmon mode traveling on a metal-dielectric interface, the wavevector in dielectric medium, in its general form, can be written as

$$\begin{aligned} \boldsymbol{k_{sp}} &= k_x \boldsymbol{a_x} + k_y \boldsymbol{a_y} + k_z \boldsymbol{a_z} \\ &= \alpha_d \boldsymbol{a_x} + i\beta \sin\theta \boldsymbol{a_y} + i\beta \cos\theta \boldsymbol{a_z}, \end{aligned} \tag{1}$$

where θ is the angle of the wavevector with the z axis, providing the metal-dielectric interface is in the yz plane. Also, α_d and β, resulted from the solution of Helmholtz wave equation on the metal-dielectric interface, are

$$\beta = k_0 \sqrt{\frac{\epsilon_m \epsilon_d}{\epsilon_m + \epsilon_d}}, \tag{2}$$

$$k_x = \alpha_d = \sqrt{\beta^2 - \epsilon_d k_0^2}. \tag{3}$$

In these equations k_0 is the wavenumber of light in free space and ϵ_d and ϵ_m are the dielectric constants of dielectric and metal, respectively. Complex dielectric constant of the metal at optical wavelengths results in a complex wavevector for surface plasmon mode and, therefore, constant amplitude and constant phase planes which are not parallel to each other. This is the characteristic of an inhomogeneous plane wave.

Figures 1(a) and 1(b) schematically show the electric and magnetic field components before reaching a dielectric lens. The boundary conditions of the metal-dielectric interface eliminate the existence of transverse electric (TE) modes and therefore only transverse magnetic or TM modes are considered here. As it is assumed in Fig. 1 the surface plasmon mode travels parallel to the optical axis before reaching the lens on its front plane, U and thus $\theta = 0$ in Eq. (1). Dielectric lenses in plasmonic regime cannot be considered as thin lenses. As a result, the phase change that is imposed by the transmission of mode through the lens should be considered in calculation of the diffracted fields behind it.

Figure 1(c) demonstrates the schematic drawing of a thick lens and a ray passing through the lens from the front plane (U) to the back plane (U'). The total phase change, $\Delta\phi$, can be found by calculating the phase change between points P_0 and P_3,

$$\Delta\phi = \beta \Delta z_1 + \beta^l \frac{\Delta z_2}{\cos\theta_1} + \beta \frac{\Delta z_3}{\cos\theta_2}. \tag{4}$$

As mentioned before, the propagation constant of the surface plasmon mode on the metal-dielectric interface before and after passing through the lens is β, but inside the lens, the value

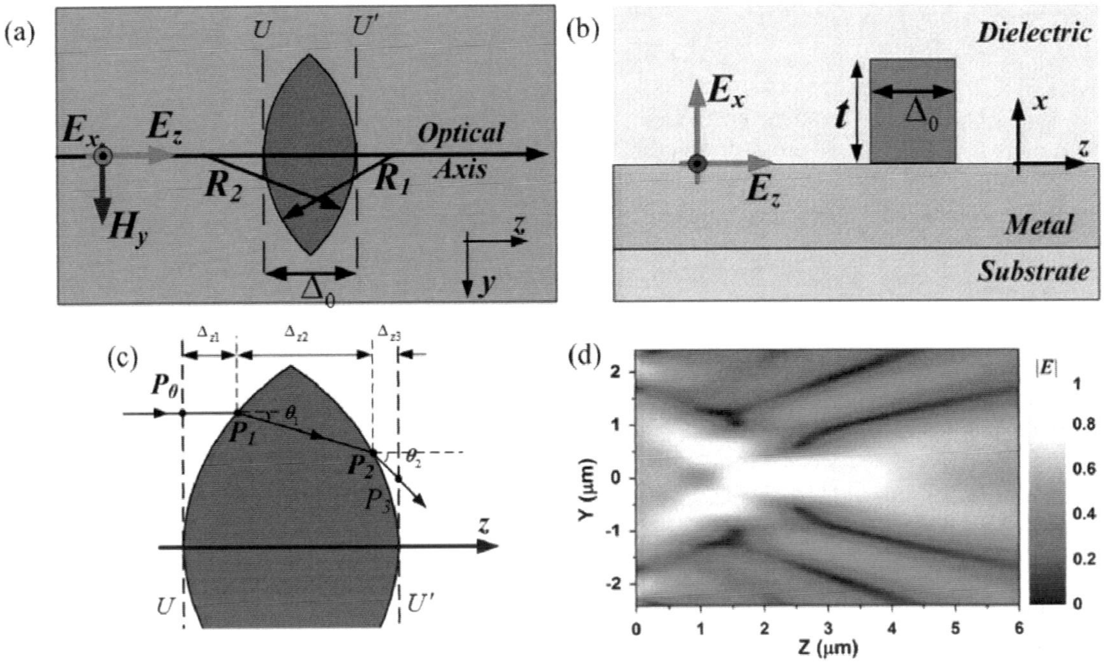

Figure 1. (a) and (b) Schematic drawing of a dielectric lens located on a metal-dielectric interface and its radii of curvatures and thickness, (c) the path of a ray traveling parallel to the optical axis as it goes through the lens, (d) magnitude of the normalized electric field behind a thick lens of radius 5 μm.

of the propagation constant changes to β^l. Δz_j, where $j = 1$, 2 or 3, denotes the distance along the z axis between successive points P_j and P_{j+1}. The angles θ_1 and θ_2 are obtained by using the Snell's law at the curved interfaces of the lens [11]

$$\theta_1 = \sin^{-1}(\frac{y_0}{R_1}) + \sin^{-1}(\frac{\beta y_0}{\beta^1 R_1}), \tag{5}$$

$$\theta_2 = \sin^{-1}(\frac{\beta}{\beta^l}\sin(\theta_1 - \sin^{-1}(\frac{y_3}{R_2}))) + \sin^{-1}(\frac{y_3}{R_2}), \tag{6}$$

where y_0 and y_3 are the y coordinates of P_0 and P_3, respectively. Using these equations, the phase delay at the backplane of the lens is obtained as a function of the height from the optical axis.

A vectorial derivation for the diffraction of surface plasmon modes is provided in [12] where the authors have used the surface wave expansion with complex wavevectors. Based on this derivation, they have presented the vectorial Huygens-Fresnel principle for the propagation of surface plasmons on a metal-dielectric interface. Their result is used here to study the propagation of surface plasmon modes behind the dielectric thick lens. After passing through the lens, the surface plasmon field in the dielectric ($x > 0$) can be represented as

$$\boldsymbol{E}(y,z) = \frac{1}{2\pi}\int \boldsymbol{E}(k_y)e^{i\sqrt{\beta^2-k_y^2}z}e^{ik_y y}dk_y. \tag{7}$$

Using the concept of the Fourier transform and convolution integral, this equation can be replaced by

$$\boldsymbol{E}(y,z) = \int E_{U',x}(y', z = \frac{\Delta_0}{2})\boldsymbol{G}(y - y', z)dy', \tag{8}$$

IOP Conf. Series: Journal of Physics: Conf. Series **961** (2017) 012004 doi:10.1088/1742-6596/961/1/012004

where

$$E_{U',x}(y', z = \frac{\Delta_0}{2}) = E_{0x} e^{-i\Delta\phi(y')}, \tag{9}$$

and E_{0x} is the amplitude of the surface plasmon in the x direction at the front plane. The other components of the electric field of the TM wave can be calculated using Maxwell's equations. $\Delta\phi(y')$ is the phase delay imposed by the lens as a function of the distance from the optical axis on the backplane. Also, in Eq. (8), $\boldsymbol{G}$ can be calculated as [13]

$$\boldsymbol{G}(y - y', z) = -\frac{1}{2}\left(\frac{\partial}{\partial z}\boldsymbol{a_x} + \frac{k_x}{\beta^2}\frac{\partial^2}{\partial z \partial y}\boldsymbol{a_y} + \frac{k_x}{\beta^2}\frac{\partial^2}{\partial z^2}\boldsymbol{a_z}\right) H_0^{(1)}\left(\beta\sqrt{z^2 + (y - y')^2}\right). \tag{10}$$

Using the equations provided, the diffraction of a surface plasmon mode is calculated as it passes through a convex lens. The normalized magnitude of the electric fields behind the lens is shown in Fig. 1(d) which is in consistence with numerical simulations. The lens is assumed to have similar radius of curvature on both sides equal to 5 µm and a thickness of $\Delta_0 = 1.75$ µm. The lens is made of PMMA with a refractive index of 1.48 and it is placed on the surface of silver. The calculations are done at the wavelength of 780 nm.

3. Higher order modes behind a dielectric lens

In solving the Helmholtz wave equations on a metal dielectric interface (the yz plane in Fig. 1) the same trend as slab waveguides is usually used and the changes in the y direction are ignored. Hence the assumption of $\partial/\partial y = 0$ results in only one TM polarized mode which is the fundamental mode of the structure, TM_0. However, without this uniformity assumption, Helmholtz equations reveal a new set of surface modes. Assuming non-uniformity in the y detection means that wavevector of surface plasmons should have a component in this direction which satisfies the dispersion relation in metal and dielectric, respectively as [14]

$$\epsilon_d k_0^2 + k_{xd}^2 - k_y^2 - \beta^2 = 0, \tag{11}$$
$$\epsilon_m k_0^2 + k_{xm}^2 - k_y^2 - \beta^2 = 0. \tag{12}$$

Using the metal-dielectric boundary conditions, it can be shown that these new set of modes have the same propagation constant in the direction normal to the surface (x axis) and therefore the same confinement in this direction. The propagation constant in the y direction, is chosen from a continuous spectrum starting from zero for TM_0 and therefore these modes are higher order plasmonic modes of the metal-dielectric interface. Higher order TM modes form an orthogonal set and therefore any linear combination of them is also a solution of the Helmholtz equation

$$E_{zd} = \int E_z(k_y) e^{-k_x x} e^{-ik_y y} e^{-i\beta z} dk_y. \tag{13}$$

The resulted fields from the diffraction of surface plasmon on a dielectric lens have a similar form with higher order surface plasmon modes when written as

$$\boldsymbol{E}(\boldsymbol{r}) = \frac{1}{2\pi}\int \boldsymbol{E}(k_y) e^{i\sqrt{K^2 - k_y^2}z} e^{-ik_y y} e^{-ik_x x} dk_y, \tag{14}$$

where K is

$$K^2 = \epsilon_d k_0^2 + k_z^2. \tag{15}$$

This shows that passing through a dielectric lens has excited higher order modes.

Although all the TM modes have similar mathematical formulation, they differ in their energy flow directions. The orientation of Poynting vector is schematically shown in Fig. 2. As θ, the

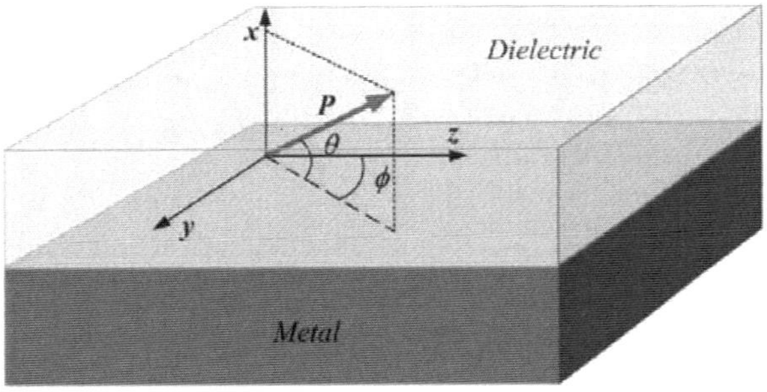

Figure 2. Schematic diagram of the Poynting vector. The dashed line shows the trajectory of P on the xy plane.

angle of Poynting vector (P) to the surface, increases, the wave behaves more like a radiation mode, transmitting energy away from the surface. On the other hand, modes having smaller values of θ are more bounded to the surface. In general, the energy flow direction of a TM mode is not in the xz plane and the trajectory of its Poynting vector on the yz plane makes an angle of ϕ with the z axis.

Here, we are concerned with the energy loss of the surface and so only focused on θ. The results of the calculation of the Poynting vector orientations are demonstrated in Fig. 3 for different points on a metal-dielectric interface. Using the vectorial Huygens-Fresnel principal and accurate SP phase delay, we have shown that higher order SP modes are excited behind a lens in an in-plane focusing system. This is equivalent to more energy loss from the surface since higher order modes are less bounded to the surface than the fundamental mode. Compared to the fundamental plasmonic mode, higher order modes are less confined to the surface and the orientation of Poynting vector can reach up to 90° from the surface in both metal and dielectric which results in extra energy loss from the surface.

4. Conclusions

Although scalar diffraction theory presents a good qualitative description of the diffraction phenomena, it fails to predict the behavior of a thick lens in plasmonic regime. In this paper, equations based on vectorial Huygens-Fresnel theory are used to calculate the phase change imposed by the lens, which can provide more accurate results of the diffracted fields. Analyzing

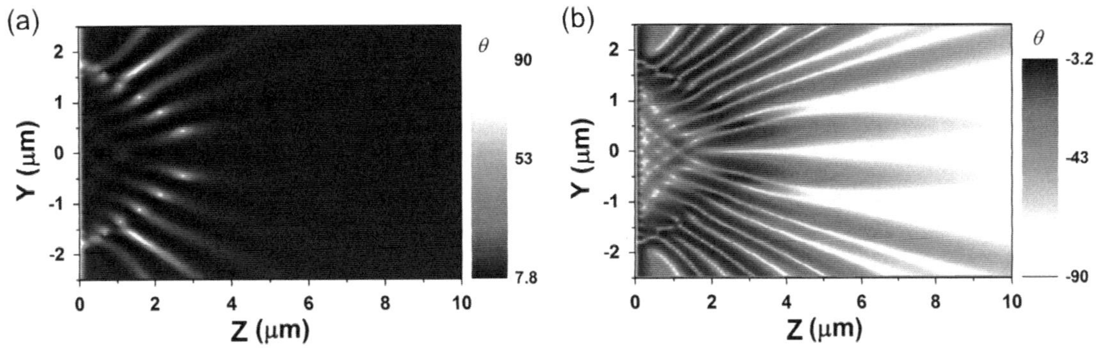

Figure 3. The energy flow of surface plasmon modes behind a dielectric lens of radius 5 μm in (a) air and (b) silver.

IOP Conf. Series: Journal of Physics: Conf. Series **961** (2017) 012004 doi:10.1088/1742-6596/961/1/012004

the obtained equations proves the excitation of higher order plasmon modes behind a dielectric lens placed on the metal-dielectric interface. Excitation of these modes would result in more energy losses since higher order modes are less bounded to the interface. We have monitored these losses from the surface in both dielectric and metal by calculating the orientation of the Poynting vector in different points behind the lens.

References

[1] Ritchie R H 1957 *Phys. Rev.* **106** 874–881
[2] Maier S A and Atwater H A 2005 *J. Appl. Phys.* **98** 011101
[3] Vincenzo A, Roberto P, Marco F, Onofrio M M and Maria Antonia I 2017 *J. Phys. Condens. Matter* **29** 203002
[4] Zia R and Brongersma M L 2007 *Nat. Nano* **2** 426–429
[5] Hohenau A, Krenn J R, Stepanov A L, Drezet A, Ditlbacher H, Steinberger B, Leitner A and Aussenegg F R 2005 *Opt. Lett.* **30** 893–895
[6] Shi W B, Chen T Y, Jing H, Peng R W and Wang M 2017 *Opt. Express* **25** 5772–5780
[7] Bezus E A, Doskolovich L L, Kazanskiy N L, Soifer V A and Kharitonov S I 2010 *J. Opt.* **12** 015001
[8] Polo J A and Lakhtakia A 2011 *Laser Photon. Rev.* **5** 234–246
[9] Oulton R F, Pile D F P, Liu Y and Zhang X 2007 *Phys. Rev. B* **76** 035408
[10] Kim H, Hahn J and Lee B 2008 *Opt. Express* **16** 3049–3057
[11] Saleh B E A and Teich M C 1991 *Fundamentals of photonics* (New York: Wiley)
[12] Archambault A, Teperik T V, Marquier F and Greffet J J 2009 *Phys. Rev. B* **79** 195414
[13] Teperik T V, Archambault A, Marquier F and Greffet J J 2009 *Opt. Express* **17** 17483–17490
[14] Kordi M, Armin F, Malekfar M R, Mirsalehi M M and Shokooh-Saremi M 2017 *Phys. Rev. A* **95** 033824

IOP Conf. Series: Journal of Physics: Conf. Series **961** (2017) 012005 doi:10.1088/1742-6596/961/1/012005

Optical perturbation of atoms in weak localization

A Yedjour[1]

[1] Faculté de Physique, USTO-MB, B.p.1505 El M'Naour, 31000.Oran, Algérie.

afifa.yedjour@univ-usto.dz

Abstract. We determine the microscopic transport parameters that are necessary to describe the diffusion process of the atomic gas in optical speckle. We use the self-consistent theory to calculate the self-energy of the atomic gas. We compute the spectral function numerically by an average over disorder realizations in terms of the Greens function. We focus mainly on the behaviour of the energy distribution of the atoms to estimate a correction to the mobility edge. Our results show that the energy distribution of the atoms locates the mobility edge position under the disorder amplitude. This behaviour changes for each disorder parameter. We conclude that the disorder amplitude potential induced modification of the energy distribution of the atoms that plays a major role for the prediction of the mobility edge.

1. Introduction

Disorder has long proven to be a crucial ingredient to understand transport properties for many different kinds of waves: with electromagnetic waves, ultrasonic waves and atomic waves [1-3]. Optical potentials used to produce disorder or confine condensates in particular directions of space offer an unprecedented experimental control. Bose-Einstein condensates thus appear to be ideal systems to study the interplay between disorder and interactions. This is reason why they are extensively used to revisit many problems of condensed-matter physics.

Weak localization leads to an enhanced probability to return back to the origin. It therefore slows down transport, i.e. it reduces the diffusion constant and the conductivity compared to the Boltzmann/Drude incoherent contribution. This effect in condensed matter systems have been widely studied theoretically and experimentally, and signatures have been found in various systems (see the reviews [4-6]). For instance weak localization correction to the conductivity has been studied as a function of temperature (which activates sources of decoherence such as phonons) and magnetic field (which introduces spin-orbit couplings and additional phases to the paths). They have also been studied as a function of disorder parameters, such as impurity concentration for example. The reviews in Ref. [5] present the dependence of the resistance of a thin disordered Mg-film, with magnetic field for several values of the temperature.

In the weak localization regime, the transport is reduced by interference between multiple scattering paths, but particles can still propagate to infinity [6]. However, in our simple model we only considered interference between loop paths that are travelled in one direction or the opposite, but interference effects are to be taken into account for many other pairs of paths, in principle all those which visit the same set of scatterer in particular those which include loops inserted in loops etc... They can lead to the complete suppression of transport, an effect which is called Anderson (or strong) localization. The localized and extended regions of the spectrum are separated by critical energies. The link between critical energy and the statistical properties of disorder has been studied in experiments with cold atoms in random optical [7- 9]. The wave propagation also is governed by scattering from the random impurities. Three typical energy-dependent length scales can be identified, which characterize three basic effects induced by the disorder. First, single scattering from impurities depletes the k-wave states, which can be seen as quasiparticles in the disordered medium, with a finite life-time$\tau_s(k)$. Single scattering hence defines the first length scale, namely the scattering mean-free path, $\ell_s = v\,\tau_s$ which characterizes the typical length travelled by the wave before it is looses the memory of its initial state. Then, multiple scattering defines the second length scale, namely the Boltzmann transport mean-free path,ℓ_B, which characterizes the typical length travelled by the wave before it looses the memory of its initial direction. In general, several scattering events are necessary to significantly deflect the trajectories so that$\ell_B, \geq \ell_s$. The two length scales are found to be equal only in the white-noise limit if it exists, where the wavelength is smaller than the typical size of the impurities [10]. The disorder is then equivalent to a set of randomly distributed Dirac peaks and the scattering is isotropic. In this case the wave looses the memory of its initial state and initial propagation direction at the same time. In the general case, within the distanceℓ_B,, the transport crosses over from ballistic to diffusive. The average squared size of the wave packet increases linearly in time,$\langle r^2 \rangle = 2dD_B t$, with $D_B = v\ell_B/d$ (where d is the space dimension) and D_B is the Boltzmann diffusion constant [11, 12].

The self-consistent theory, first formulated by Vollhardt and Wolfle in 1981 for 2D electron conductivity [13, 14], was the first work that explicitly calculated quantum corrections to the classical"Drude"-conductance. Beyond this theory, the diffusion of the BEC in a random potential has been studied [15–18] and it was pointed out that only some fraction of the condensate diffuses away, while the rest stays localized near its initial location (Anderson transition for BEC) [17, 18].
The macroscopic transport properties of a coherent wave in a disordered medium, namelydiffusion and weak/strong localization, closely depend on the microscopic and statistical properties of the disorder itself. In this paper we study quantum transport of matterwaves in disordered potentials with isotropic correlations in 3d. The isotropy provides a great simplification, and other methods, such as the self-consistent Born approximation, can be used to calculate the self- energy. Many experiments reported the observation of an Anderson transition in momentum space using cold-atom kick-rotor setups [19, 20], study of classical diffusion in two-dimensional (2D) speckle potentials [21] which is revelant for current experiments on ultra-cold atoms in speckle disorder. The calculation of any measurable quantity is specific to the particular realization of the disorder. Therefore, meaningful quantities correspond to statistical averages over realizations of disordered potentials. When averaging over disorder realizations, some quantities can be written in terms of the average Green function$G(E)$it is the case the spectral function $A(E, \boldsymbol{k})$defined by [22,23]. It contains all the information about the spectrum of the disordered medium. For a particle in disorder-free space, it is given by $A(E, k = 2\pi\langle\delta\big(E - \varepsilon(k)\big)\rangle$with$\varepsilon(k) = \frac{\hbar k^2}{2m}$. The average density of states (per unit volume) reads:$N(E) = \int \frac{dk}{2\pi^d}\frac{A(E,\boldsymbol{k})}{2\pi}$.

This paper uses the first Born Approximation to study the behavior of the energy distribution of the atoms in an optical specklefor two values of disorder amplitude. We expect finding the true mobility edge and possibly leading to the Anderson localization phenomenon.

2.Model

We Consider a weakly interacting Bose-Einstein condensate $of\ N \gg 1$ atoms of mass m expanding in a three-dimensional random potential$V\,(r)$. The origin of time is chosen within the third stage of the toy model described inRef [24] which ensures that interactions are neglect. We calculate the density profile of the condensate inserting the expression of the random potential see in Ref [25]. The potential disorder obtained inside an integrating sphere lit with a laser beam, the real-space correlation function of which reads $C(r) = V_R^2 \frac{(\sin r)^2}{r^2}$with r normalized by σ. σ:is the correlation length the correlation [24].

The spectral function permits to relate the energy distribution $\varphi_E$and the momentum distribution $\varphi_k$of the stationary particles in the disorder: $\varphi_E = \int \frac{dk}{2\pi^d} A(E, \boldsymbol{k})\varphi_k$ which is normalized by $\int dE \, \frac{A(E,\boldsymbol{k})}{2\pi} = 1$. The calculation of the energy distribution therefore requires the knowledge of the full spectral function, i.e. of the real and imaginary parts of the self -energy see in this equation:

$A(E, \boldsymbol{k}) = (-Im\Sigma(\varepsilon, k))/(E - \varepsilon(k) - Re\Sigma(\varepsilon, k))^2 \; + \; (\mathrm{Im}\Sigma(\varepsilon, k))^2$

With$Re\Sigma(\varepsilon, k) Im\Sigma(\varepsilon, k)$) are the real and imaginary parts of the self-energy Σrespectively.The quantity $Re\Sigma(\varepsilon, k)$ thus describes the shift in energy of the free-particle modes when they are dressed by the disorder. The quantity $\mathrm{Im}\Sigma(\varepsilon, k)$ is the energy width of the spectral function, which defines the scattering mean free time $\tau_s = -\frac{\hbar}{2\mathrm{Im}\Sigma(\varepsilon,k)}$. Therefore the spectral function contains all the information about the relative weight, the energy, and the lifetime of the quasi-particles, i.e. the particles dressed by the disordered medium, which on average define an effective medium.

The microscopic parameters in these systems are precisely known and in many cases, which paves the way to direct comparison between experiments and theory [25]. This isaGreat advantage of ultracold atoms, compared to traditional condensed matter systems. MajorAdvances reported so far are the observation of one-dimensional (1D) Anderson localization ofMatterwaves [26, 27], and studies of the effects of weak [28-31] and strong [32] interactions in disordered gases. All those studies have muchbenefited from the close interplay between theory and experiments.

The aim of thins work is to discussthe transport and localization properties of matter waves in 3D isotropic optical speckle potentials. We will focus on the mobility edge ε_c, which is the solution of $E - \varepsilon(k) - Re\Sigma(\varepsilon, k) - Im\Sigma(\varepsilon, k) = 0$.Here, we incorporate$Re\Sigma(\varepsilon, k)$ and $Im\Sigma(\varepsilon, k)$)into the spectral function in first approximation.

3.Results and discussion

In this section, we discuss the numerical results on the distribution of energy expressed fromthe self-energyby using Born approximation. The angular integral of the self-energy$\Sigma(\varepsilon,k)$ was calculated analytically for first Born approximation[32]and we also calculate it numerically.

figure 1 shows the imaginary part of $\Sigma(\varepsilon, k)$for three values of the energyε.The curve shows that the behavior of $Im\Sigma(\varepsilon, k)$is different for three values of ε and its clearly negative. This shifts the band edge of the energy spectrum to $\Delta\varepsilon = 0.2$(normalized by ε_σ).

Weobserved that$Im\Sigma(\varepsilon, k)$is nonzero only for$0 \leq \varepsilon \leq$. 2.78atε=0.5 comparing with the other two values. This behavior assumes that the number of atoms expanding in therandom disorder increases and it can be seen that the first Born approximation significantly overestimates the amount of scattering in this case. This behavior is qualitatively similar to that obtained in Ref [33]butwith$\varepsilon = 1$ with being another consequence of long-range correlations.

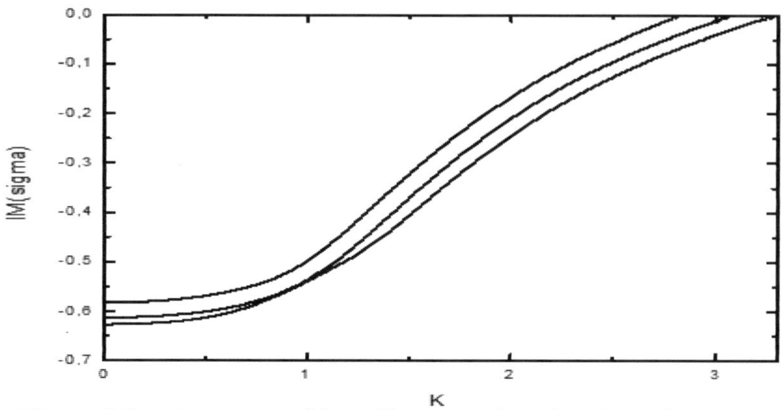

Figure 1. Imaginary part of the self- energy (in units of ε_ξ) of an atom in a speckle potential with $\varepsilon=0.5, 1 and 1.5$, as a function of k in first Born approximation.

To find the spectral $A(\varepsilon,k)$ we go to beyond the for first Born approximation [20, 26] at energyε. At larger energies $\varepsilon>3$, the spectral function behaves as a free space expression $A(\varepsilon,k) \propto \delta(\varepsilon- \varepsilon_k)$ strongly peaked near$k=\sqrt{\varepsilon}$[24].

For a short-range correlations, the self-energy is independent of k and the spectral function has a large energy tail ($A(\varepsilon,k) \propto 1/\varepsilon^{3/2}$) and atoms with high energies exhibit diffusion behavior ($\mu >> \varepsilon$) [21]. Thus, a potential with correlation reduces the large energy tail of $A(\varepsilon,k)$ and seems to help in localizing more atoms.

We go beyond the for first Born approximation to solve numerically the spectral function $A(\varepsilon,k)$ for$V_R = 1$ and $V_R = 2$. The curve observed in figure 2shows that the edges of the spectrum are different, one with positive energy, the atom expends from zero to 2. Other one withpositive energy exhibits a peak with small energy width. This peak related to the real part of the self-energy assume us that$\int A(\varepsilon,k)\, d\varepsilon= 1$ with a constant amplitude of disorder.

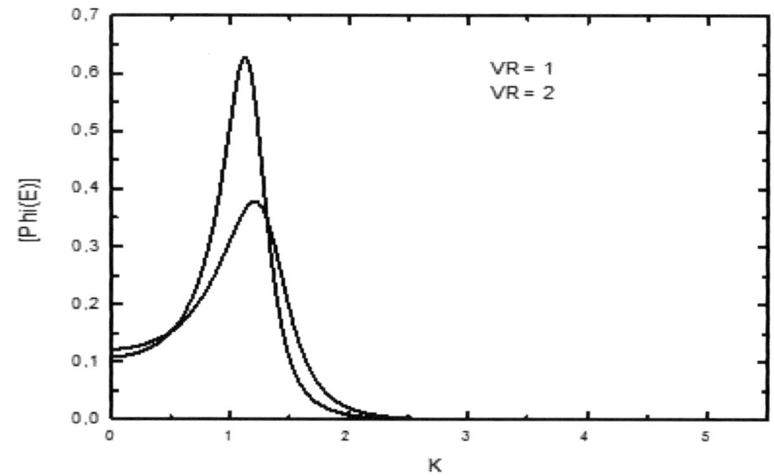

Figure 2. The distribution of energy expressed from the self enery by using Born approximation at $\varepsilon=1$, for $V_R = 1$ and $V_R = 2$

We solve by iterations the distribution of energy expressed from spectral function. The energy spectrum becomes continuous and larger. It has a large weight and extends up to ε= 2, many atoms achieve energies ε. These results are studied with predictions based in Ref [21] whose found 45%

atoms localized for $V_R = 1$at small chemical potential.In figure 3, we observed the distribution of energy $N(E)$which is the probability to find an atom with energyεbut on a logarithmic scale. In the presence of the random potential, one find at $\varepsilon= 0$ the probability remains finite and nonzero.and they obtain 35% atoms delocalized [24].

Our numerical results show that the first Born approximation locates the band edge at larger energies and extends up $\varepsilon=2$.The amplitude disorder takes an important place on energy spectrum when we include the first order corrections in Born approximation. We expect that the localization occurs near the mobility edge which is locates at$\varepsilon = -0.34$.

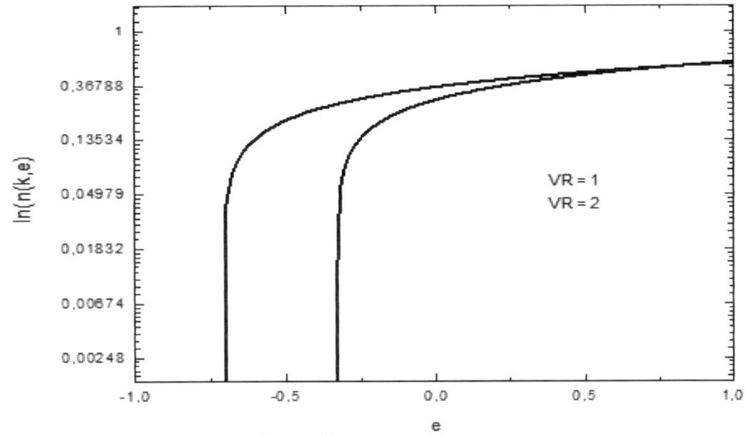

Figure3. $\ln(N(e))$per unit volume, for $K = 1$

We predict the possible presence of a mobility edge in this region. We observe that the disorder amplitude provided by the optical disordered potential causes a shift of the band edge towards the negative energy.

This work may expect to find thetrue mobility edge position ina particular way to measure the exact value of ε_c, and to determine critical exponents for localization.

4.Conclusion

In this work, we discuss the optical perturbation of atoms in weak localizationfor the first order Born approximation. We calculate imaginary part of the self-energy for different value of the disorder amplitude to study the behavior of self -energy on the distribution of energy. The self-energy $\sum(\varepsilon,k)$ is a complex function, we found that its imaginary part shifts the band edge of the energy spectrum. This part of the self-energy has an important role on the energies distribution. In the first order Born approximation, the energy spectrum of the spectral function has extend from$\varepsilon= 0$, to $\varepsilon= 2$, it can be seen that the first order Born approximation significantly overestimates the amount of scattering in this case. We predict a possible presence of the mobility edge in this region. We observe that the disorder amplitude provided by the optical disordered potential changes the form of the curve and shifts the band edge to a negative energy. Also, we examine the behavior of the spectral function for two values of constant disorder amplitude to locate the mobility edge position, we find out some remarkable differences between them. We assumes that the number of atoms expanding in the random disorder increases in this case the energy spectrum becomes continuous with a maximum equal to 0.64 for $V_R = 2$.The behavior of the density of states is different compared to the density of states of free particles. By this way we have plotted the probability of distribution of energy at logarithmic scale to locate the mobility edge. We find that the disorder amplitude provided by the optical speckle is responsible for the position of the mobility edge.

"I thank Bart Van Tiggelin for his help to work on the cold atoms"

IOP Conf. Series: Journal of Physics: Conf. Series **961** (2017) 012005 doi:10.1088/1742-6596/961/1/012005

References

[1] A. Lagendijk, B.A. van Tiggelen and D.S. Wiersma. (2009) Physics Today **62**, 24.
[2] A. Aubry, L. A. Cobus, S.E.Skipetrov, B. van Tiggelen, A. Derode, and J.H.Page. (2014) Phys. Rev. Lett. **112**, 043903.
[3] A. Aspect and M. Inguscio, (2009) Physics Today **62**, 30.
[4] P. A. Lee and T. V. Ramakrishnan. (1985) Rev. Mod.Phy. **57,** 287–337.
[5] G. Bergmann. (1984). Phys. Rep. **107**,1.
[6] M. E. Gershenson B. L. Altshuler, A. G. Aronov and Yu. V. Sharvin. (1987) Phys. Rev. Sov. Sci. Rev. **9**, 225.
[7] M. Piraud, P. Lugan, P. Bouyer, A. Aspect and L. Sanchez-Palencia. (2011) Phys. Rev. A **83,** 031603
[8] S. S. Kondov, W. R. McGehee, and B. Zirbel, J. J. DeMarco, (2011) Science **66**, 334.
[9] F. Jendrzejewski, A. Bernard, K. Mller, P. Cheinet, V. Josse, M. Piraud, L. Pezz, L. Sanchez-Palencia, A. Aspect, and P. Bouyer. (2012) Nature Phys **8**, 398.
[10] S.E. Skipetrov, A. Minguzzi, B.A. van Tiggelen and B. Shapiro. (2008) Phys. Rev. Lett. **100,** 165301.
[11] Neil W. Ashcroft and N. David Mermin, (1976).Solid State Physics Thomson Learning,New York.
[12] Nevill F. Mott. (1990). Taylor & Francis, London, UK
[13] A. Lagendijk and B.A. van Tiggelen. (1996) Phys. Rep. **270,** 143.
[14] B. Shapiro. (2012) J. Phys. A: Math. Theor **45**, 143001.
[15] B. Allard, T. Plisson, M. Holzmann, G. Salomon, A. Aspect, P. Bouyer, and T. Bourdel. (2012) Phys. Rev. A **85**, 033602.
[16] M. C. Beeler, M. E. W. Reed, T. Hong, and S. L. Rolston. (2012) New J. Phys. **14,** 073024.
[17] L. Pezz and L. Sanchez-Palencia. (2011) Phys. Rev. Lett., **106**, 040601.
[18] D. Vollhardt and P. Wolffle; W. Hanke and Ya. V. Kopaev (1992). Eds, (North-Holland, Amsterdam,
[19] J. Chabe, G. Lemarie, B. Gremaud, D. Delande, P. Szriftgiser, and J.-C Garreau. (2008). Phys. Rev. Lett. **101**,255702.
[20] M. Lopez, J.-F. Clement, P. Szriftgiser, J.-C. Garreau, and D. Delande. (2012) Phys. Rev. Lett. **108**, 095701.
[21] L. Pezze, M. Robert-de-Saint-Vincent, T. Bourdel, J.-P. Brantut, B. Allard, T. Plisson, A. Aspect, P. Bouyer, and L. Sanchez-Palencia. (2011) New J.Phys. **13**, 095015.
[22] Jorgen Rammer. .(1998). Perseus Books, Reading, Mass
[23] R.C. Kuhn, C. Miniatura, O. Sigwarth,D. Delande, and C.M. Muller.(2007) New J. Phys. **9,** 161.
[24] A. Yedjour and B. A. van Tiggelen. (2010) Eur. Phys. J. D, **59** 249255.
[25] M. Piraud, P. Lugan, P. Bouyer, A. Aspect, and L. Sanchez-Palencia. (2011) Phys. Rev. A, **83** 031603(R),.
[26] J. Billy, V. Josse, Z. Zuo, A. Bernard, B. Hambrecht, P. Lugan, D. Clement, L. Sanchez-Palencia, P. Bouyer, and A. Aspect. (2008) Nature (London) **453,** 891–894,
[27] G. Roati, C. D'Errico, L. Fallani, M. Fattori, C. Fort, M. Zaccanti, G. Modugno, M. Modugno, and M. Inguscio. (2008) Nature (London), **453** 895–898,.
[28] G. Kopidakis, S. Komineas, S. Flach, and S. Aubry (2008). Phys. Rev. Lett., **100** 084103,
[29] T. Paul, P. Schlagheck, P. Leboeuf, and N. Pavloff. (2007) Phys. Rev. Lett., **98** 210602,.
[30] T. Paul, M. Albert, P. Schlagheck, P. Leboeuf, and N. (2009) Phys. Rev. A, **80** 033615,
[31] A,Yedjour and al. (2017) .Journal of Computational Electronics **13**,1
[32] L. Fallani, J. E. Lye, V. Guarrera, C. Fort, and M. Inguscio (2007). Phys. Rev. Lett., **98** 130404.
[33] Y. Castin and R. Dum, (1996).Phys. Rev. Lett. **77**, 5315.

IOP Conf. Series: Journal of Physics: Conf. Series **961** (2017) 012006 doi:10.1088/1742-6596/961/1/012006

Performance Enhancement of Bidirectional TWDM-PON by Rayleigh Backscattering Mitigation

Ibrahim A. Elewah[1*], Martina N. Wadie[2] and Moustafa H. Aly[3]

[1] Computer Engineering Technology, American College of the Middle East (ACM), Egaila Area, Kuwait, IEEE Member.

[2] Electrical and Computer Engineering, Higher Technological Institute (HTI), 6th of October, Egypt.

[3] Electronics and Communication, College of Engineering and Technology, Arab Academy for Science, Technology and Maritime Transport, Alexandria, Egypt, OSA Member.

*Corresponding author E-Mail: ibrahim.elewah@acm.edu.kw

Abstract. A bidirectional time wavelength division multiplexing-passive optical network (TWDM-PON) with a centralized light source (CLS) is designed and evaluated. TWDM-PON is the promising solution for PON future expansion and migration. The most important issue that limits optical fiber transmission length is the interferometric noise caused by Rayleigh backscattering (RB). In this study, we demonstrate a TWDM-PON architecture with subcarrier at the remote node (RN) to mitigate the RB effect. A successful transmission with 8 optical channels is achieved using wavelength division multiplexing (WDM). Each optical channel is splitted into 8 time slots to achieve TWDM. The proposed scheme is operated over 20 km bidirectional single mode fiber (SMF). The proposed system has the advantage of expanding the downstream (DS) capacity to be 160 Gb/s (8 channels×20 Gb/s) and 20 Gb/s (8 channels×2.5 Gb/s) for the upstream (US) transmission capacity. This is accomplished by a remarkable bit error rate (BER) and low complexity.

1. Introduction

Recently the time and wavelength division multiplexed passive optical network (TWDM-PON) has been chosen by the Full Service Access Network (FSAN) as a primary solution for next generation PON stage2 (NG-PON2), owing to its features of backward compatibility, components availability and technical ripeness [1, 2]. Moreover, with increasing the bandwidth due to the requirement of intensive applications, the main aim of future high capacity PON system is to prepare a high access rate for an increased number of subscribers [3].

Besides high capacity and large subscriber numbers, the important matter that is very concerned by the users is the low cost and effective system upgrade. The scheme of employing CLS to remodulate the DS signal for US transmission without power source at ONU side is the attractive solution to laser saving and to reduce complexity [4, 5]. Also, CLS schemes suffer from the cross talk that occurs due to the utilization of the same wavelength for DS and US and reflections [5]. Before remodulating DS, it has been exposed to several reflections and RB in the DS direction. These reflections are amplified and highly affect the US data [6]. Researches designed and evaluated various architectures to compensate the RB effect leading to a remarkable performance. Reduction and mitigation are the main two strategies that are used to compensate the RB [5, 7, 8].

In this paper, a bidirectional TWDM-PON scheme is proposed and evaluated to mitigate the RB interferometric noise. This is carried out using a Mach-Zehnder modulator (MZM) that works as a carrier suppressor and a subcarrier generator at RN. A successful transmission is achieved with 8 optical channels using wavelength division multiplexing (WDM). Each optical channel is splitted into 8 time slots to achieve TWDM. The proposed scheme uses differential-phase-shift-keyed (DPSK) for 20 Gb/s DS and on-off key (OOK) for 2.5 Gb/s US and operates over 20 km bidirectional single mode fiber (SMF).

The remainder of the paper is organized as follows. Section 2 is devoted for a detailed description of the proposed system. System evaluation and simulation results are presented in Sec. This is followed by the main conclusions to show the merits of this work in Section 4.

2. Proposed WDM-PON system architecture

2.1. Introduction

Fig. 1 illustrates the proposed system architecture: a full-duplex 8-channels TWDM-PON centralized light source. This architecture is capable of transmitting 20 Gbit/s DS and 2.5 Gbit/s US per channel. The eight generated downstream return-to-zero differential-phase-shift-keyed (RZ-DPSK) signals are multiplexed with a multiplexer (MUX). The multiplexed signals are transmitted over a 20 km bidirectional optical fiber.

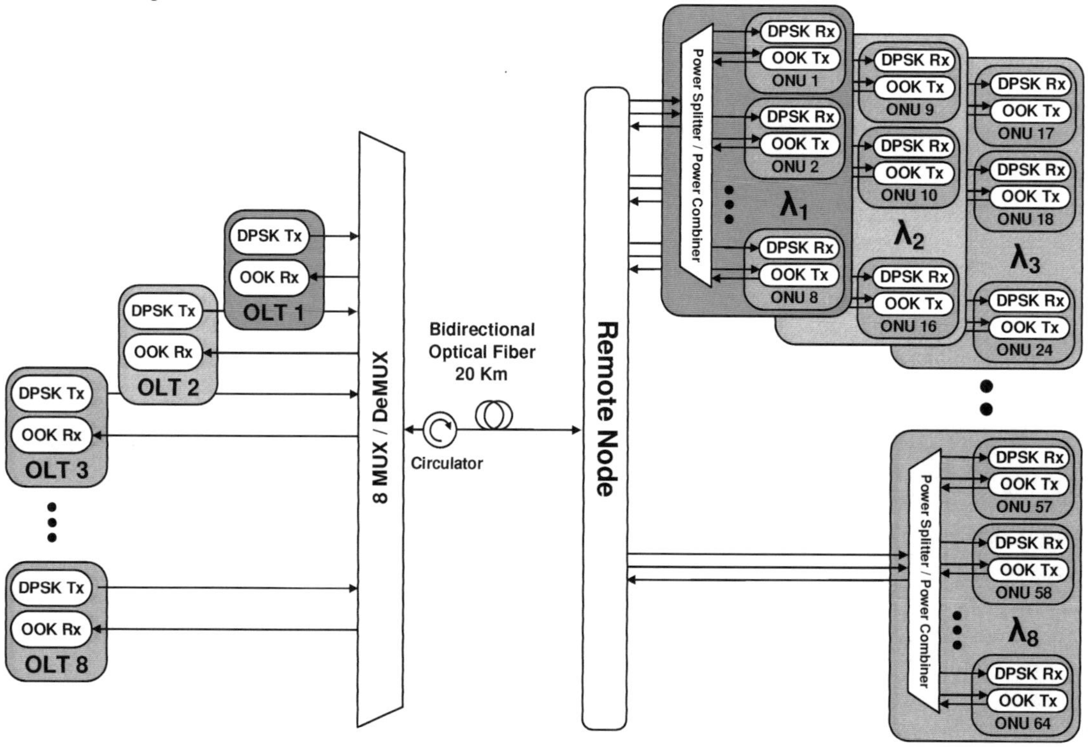

Figure 1. A full-duplex 8-channels TWDM-PON CLS.

At the RN, a MZM is used to suppress 8 carriers (8 DS signals) which contain the main power of RB and generates two subcarriers for each carrier. A demultiplexer (DeMUX) at RN is used to demultiplex the downstream signals (2×8 subcarriers). After demultiplexing 16 subcarriers, each two subcarriers (one is used to detect DS signal and the other to be remodulated for the US) send to their respective ONU. Each ONU splits the optical signal into 8 time slots. OOK modulation technique is used for US signals to retransmit to their respective OLT. Table 1 represents the operating parameters applied to the proposed system and that are extracted from literatures and standards.

Table 1. WDM-PON system parameters.

Parameter	Value	Unit
Bit Rate Downstream	20	Gb/s
Bit Rate Upstream	2.5	Gb/s
Distance [a]	20	km
Channel Spacing [b]	50	GHz
Channels Center Frequency	193.1	THz
No. of WDM Channels	8	N/A
No. of Time slots	8	N/A
Signal Power	10	dBm

[a] Distance matches the standards [9].
[b] Channel Spacing matches fixed-grid nominal central frequencies for WDM [10].

2.2. System Structure

2.2.1. DPSK Transmitter at OLT. The transmitter (Tx) consists of a continuous wave (CW) laser source with 10 dBm launched power which is supplied to one of the input of Mach-Zehnder modulator 1 (MZM1) as shown in Fig. 2. A 20-Gb/s non-return-to-zero (NRZ) pattern is generated with a duobinary precoder and is applied to the second input of MZM1 to form the DPSK. After that, the DPSK is applied to the second modulator, MZM2 (pulse carver), that cuts out only the amplitude-modulation-free center portions of the bits, thereby, largely eliminating any residual dips [11].

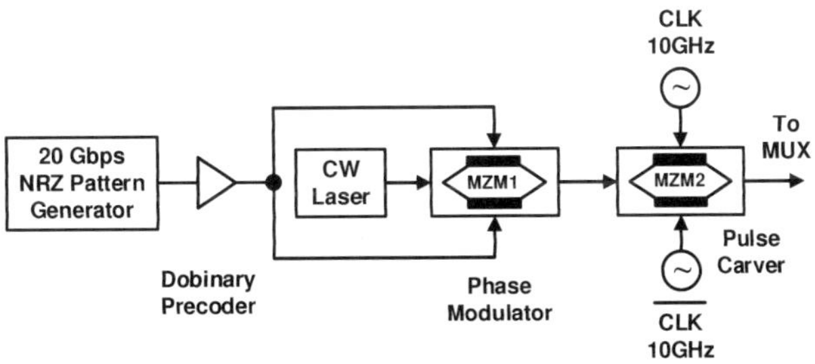

Figure 2. DPSK transmitter internal structure.

2.2.2. Bidirectional Fiber. Now, the 8 optical channels are prepared for multiplexing using MUX and are transmitted over a 20 km bidirectional single-mode fiber (SMF). The SMF specifications are tabulated in Table 2 [12, 13].

Table 2. Optical fiber parameters.

Parameter	Value	Unit
Dispersion	16.75	ps/km/nm
Dispersion Slope	0.075	$ps/nm^2/km$
RB coefficient	5e-005	km^{-1}

2.2.3. Remote Node. Figure 3 illustrates the RN structure. A MZM receives the 8 channels and the 12.5 GHz sinusoidal wave to suppress the carriers and generate two subcarriers for each carrier. A 16 subcarriers are generated with 25 GHz spacing. Then, they are demultiplexed using 1×16 DeMUX. Eight subcarriers are used to extract the DS information at DPSK receiver (Rx). The other eight subcarriers are used for US remodulation. By using one MZM at RN for the 8 channels, the US channels are shifted by 12.5 GHz from the DS channels to mitigate the RB. This is the main contribution of our proposed TWDM-PON with low complexity.

In Ref [7], the architecture uses two bidirectional fibers to mitigate the RB carrier. A DPSK modulation is utilized at ONU of architecture Ref. [8] to reduce the RB effect. In [14], a pair of electrical orthogonal codes is generated to mitigate RB.

2.2.4. DPSK Receiver at ONU. At each ONU, there are 8 DPSK receivers for the 8 times slots. Each consists of a delay interferometer (DI) with a delay (1/DS bit rate) and a coupling coefficient of 0.5. The two outputs of DI are applied into two PIN photodetectors to convert the optical signal into electrical signal. A low pass filer (LPF) is used to refine the electrical signal before subtracting. The DPSK receiver is shown in Figure 4.

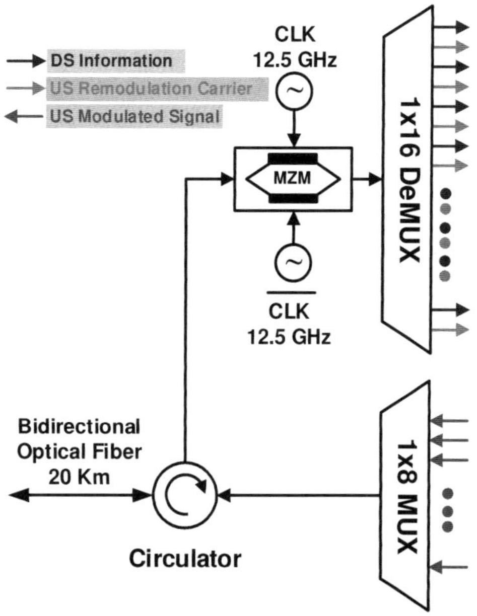

Figure 3. Remote Node internal structure.

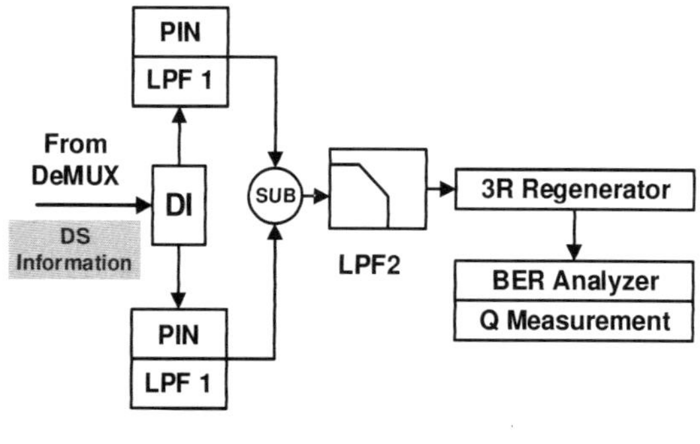

Figure 4. DPSK receiver internal structure.

2.2.5. OOK Transmitter at ONU. The OOK transmitter is a simple modulation format shown in Figure 5. A MZM modulates the US subcarrier with a 2.5-Gb/s NRZ pattern generator. Each ONU contains 8 OOK transmitters for 8 time slots. After remodulating the US signals, they are multiplexed by using MUX and are retransmitted over 20 km bidirectional fiber.

2.2.6. OOK Receiver at OLT. Each OLT receives its own optical signal, after demultiplexed by DeMUX, on PIN photodetector to convert the optical signal into electrical signal. A LPF is used to refine the electrical signal and prepare it for data extraction. Figure 6 illustrates the internal structure of the OOK receiver.

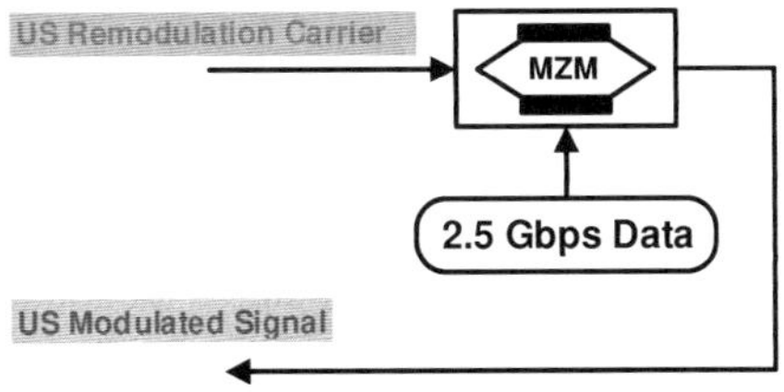

Figure 5. OOK transmitter internal structure.

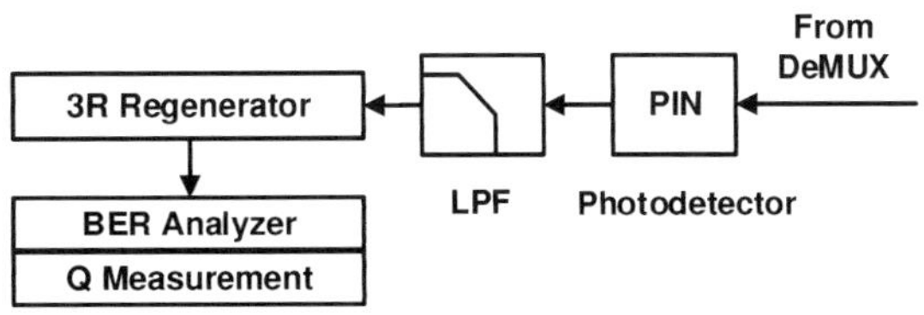

Figure 6. OOK receiver internal structure.

3. Results and analysis

In this section, the proposed TWDM-PON architecture is verified and evaluated using simulation and the obtained results are discussed. In the OLTs, eight 50-GHz-spaced wavelengths are modulated by DPSK modulation format to be multiplexed as shown Fig. 7(i). In Fig. 7(ii), RB is illustrated (the blue signal) after transmitting the DS signals over 20 km. The RB main power (~ 70 dBm) is centered at each signal. To mitigate the RB, at the RN a MZM receives the 8 channels to suppress the carriers and generate two subcarriers for each carrier. 16 subcarriers are generated with 25 GHz spacing as shown in Fig. 7(iii). This idea has the advantages that DS signal becomes free from the RB effect and the US signals are clear from RB before retransmitting as shown in Fig. 7(iv) and without using another fiber as in Ref. [15].

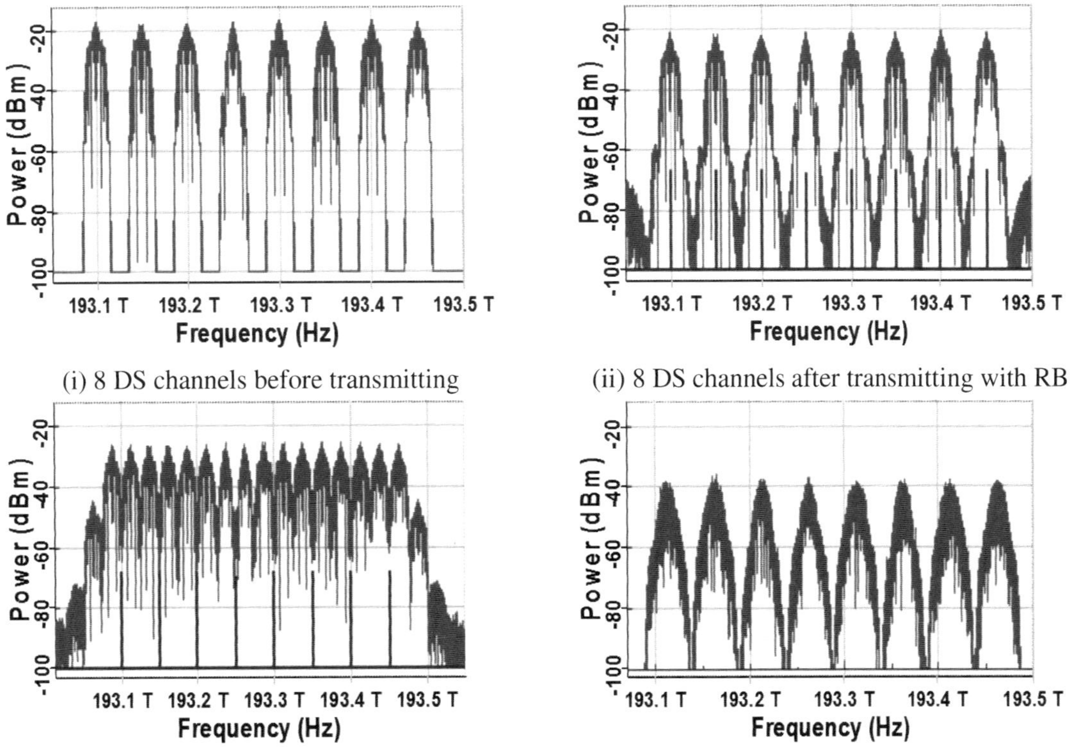

(i) 8 DS channels before transmitting

(ii) 8 DS channels after transmitting with RB

(iii) 16 subcarriers with 25 GHz spacing

(iv) 8 US channels before retransmitting.

Figure 7. Channels spectrum for each stage structure.

Figure 8 shows eye diagrams for DS and US signal. Their performances demonstrate the system feasibility and Q- factor. It is obvious that the Q- factor for both channels are greater than 7.0 which corresponds to bit-error-rate (BER) of $\sim 10^{-13}$.

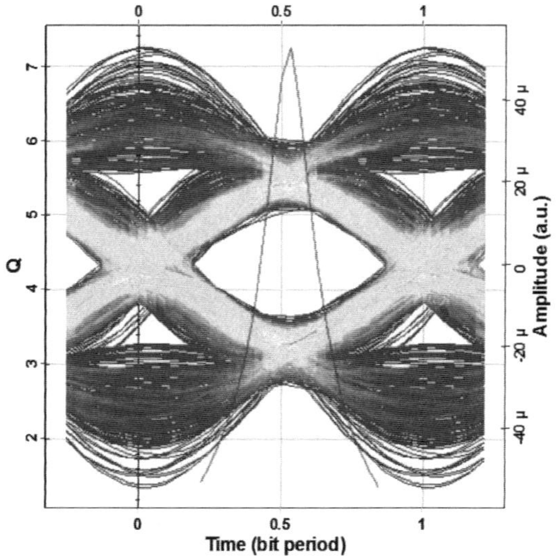

(i) DS eye diagram

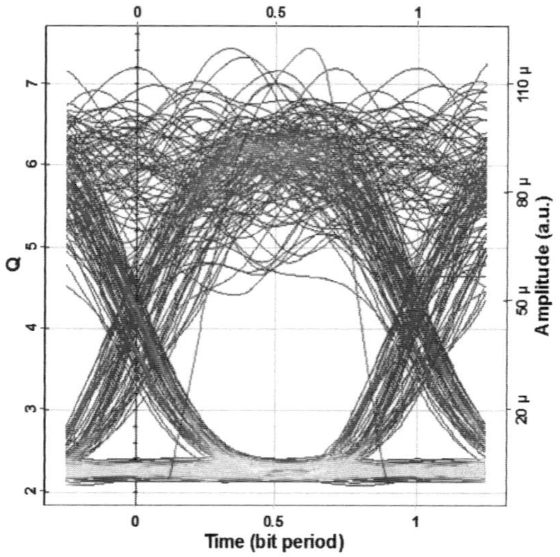

(ii) US eye diagram.

Figure 8. Eye diagrams for DS and US signal.

4. Conclusion

A bidirectional TWDM-PON scheme is proposed and evaluated to mitigate the RB interferometric noise. This is accomplished by using only one MZM at RN. A successful transmission with 160 Gb/s DS and 20 Gb/s US optical channels using WDM technique is achieved. Each optical channel is splitted into 8 time slots to achieve TWDM. This scheme is capable of serving 64 users over 20 km bidirectional SMF with BER about 10^{-13}.

References

[1] Effenberger F 2012 *38th European Conf. and Exhibition on Optical Comm. (Amsterdam)* vol. 30, no. 13, pp. 2151 - 2155
[2] Luo Y, Zhou X, Effenberger F, Yan X and Peng G 2013 *J. of Lightw. Technol.* **31** (4), 587–93
[3] Cho K , Hong U, Takushima Y and Agata A 2012 *IEEE Photonics Tech. Lett.,* **24** (3), 209–11
[4] Wang H and Chow W 2010 *IEEE Photonics Technol. Letters,* **22** (11), 820–2
[5] Elewah I, Mohammed N and Aly M 2017 *J. of Nanoelectronics and Optoelectr.,* **12** (3), 242–46
[6] Chow C, Yeh H and Xu L 2010 *IEEE Photonics Technol. Letters,* **22** (17), 1294–96
[7] Arellano C, Langer K D and Prat, J 2009 *J. of Lightw. Technol.* **27** (1), 12–8
[8] Yeh H, Chow W and Chen Y 2012 *J. of Lightw. Technol.,* **30** (13), 2151–5
[9] Xu J, Li M and Chen K 2011 *J. of Lightw.Technol.* **29** (24), 3632–9
[10] Prat, J 2008 *Next-Generation FTTH Passive Optical Networks Research Towards Unlimited Bandwidth Access* Barcelona, Spain: Springer
[11] ITU 2012 *Recommendation ITU-T G6941* The International Telecommunication Union (ITU)
[12] Gnauck A and Winzer P 2005 *IEEE J. of Lightw. Technol.* **23** (1), 115–30
[13] Liaw S, Dou L and Xu A 2007 *Optics Express,* **15** (19), 12356–61
[14] Tan Z, Wang Y, Ren W and Liu Y 2009 *Int. J. for Light and Electr. Optics,* **120** (1), 9–13
[15] Feng Q, Li W, Wang Y, Zheng Q, He Z, Yang Q and Yu S 2016 *J. of Lightw. Technol.* **34** (3), 845–53
[16] Zhou Z, Xiao S, Qi T, Li P, Bi M and Hu W 2013 *IEEE Photonics J.,* **5** (4)

IOP Conf. Series: Journal of Physics: Conf. Series **961** (2017) 012007 doi:10.1088/1742-6596/961/1/012007

Control of spontaneous emission rate in luminescent resonant diamond particles

R. Savelev[1], A. Zalogina[1], S. Kudryashov[1,2], A. Ivanova[2], A. Levchenko[2], S. Makarov[1], D. Zuev[1], I. Shadrivov[1,3]

[1]ITMO University, St. Petersburg 197101, Russia
[2]Lebedev Physical Institute, Moscow 119991, Russia
[3]Nonlinear Physics Center, Australian National University, Canberra, Australia

E-mail: r.savelev@metalab.ifmo.ru

Abstract.
We study the properties of luminescent diamond particles of different sizes (up to ~1.5 μm) containing multiple NV-centers. We theoretically predict that the average liftetime in such particles is decreased by several times as compared to optically small subwavelength nanodiamonds. In our experiments, samples were obtained by milling the plasma-enhanced chemical vapor deposited diamond film, and characterized by Raman spectroscopy and dark-field spectroscopy methods. Time-resolved luminescence measurements of the excited state of NV-centers showed that their average lifetime varies from 10 to 17 ns in different samples. By comparing this data to the values of the lifetime of the NV-centers in optically small nanodiamonds, known from literature, we confirm a severalfold decrease of the lifetime in resonant particles.

1. Introduction

Resonant high-refractive-index nanoparticles and nanostructures represent a promising platform for effective light manipulation at nanoscale and pave the way for creation of novel photonic devices. Interaction of a quantum emitter with one or several dielectric particles allows to significantly alter its radiation pattern and spontaneous emission rate [1]. Diamond particles containing colour centers such as nitrogen-vacancy, silicon-related and nickel-related centers [2] are of special interest in this context. They can be viewed as a single photonic nanodevice with emitting source embedded inside a diamond particle. Indeed, relatively high refractive index ($n \approx 2.4$) of the diamond and almost zero absorption in visible frequency range make it possible to exploit the morphology- and size-dependent resonance properties of the diamond particles to control the emission properties of luminescent centers [3], which is of particular interest for the development of quantum emitters [4, 5]. In particular, the NV-centers in diamond are of a great importance as sources of single photons, that provide the zero-phonon-line (ZPL) emission in the photoluminescence (PL) spectrum at 637 nm even at room temperature.

Usually, the NV-centers are coupled to the photonic nanostructures in order to manipulate their radiating properties and decrease the emission into phonon sidebands [6]. As an example, coupling of a single NV-center to a photonic crystal cavity fabricated in a single-crystal diamond provides 70% emission of photons with ZPL wavelength into the cavity mode [7]. However, such systems are hard to fabricate and integrate with other nanophotonic devices. On the other

IOP Conf. Series: Journal of Physics: Conf. Series **961** (2017) 012007 doi:10.1088/1742-6596/961/1/012007

hand, resonant diamond particles containing light-emitting point defects in fact represent an integral active optical element. Moreover, recent theoretical studies predict the increase of the spontaneous emission rate of quantum sources embedded within the high-index dielectric particles with the Mie modes [8, 9, 3]. Thus, the creation of active nanoantennas based on resonant nanodiamonds with embedded NV-centers is of a high importance not only for nanophotonics in general, but also for different areas of the modern science ranging from quantum information processing to optical bioimaging and sensing.

Here, we study active dielectric nanoantennas made of resonant diamond particles of different sizes up to $\sim$1.5 μm with incorporated luminescent NV-centers. We study theoretically the emission properties of NV-centers affected by resonance structure of diamond particles. We realize such nanoantennas experimentally and demonstrate the effect of the lifetime reduction of NV-centers due to the presence of resonances in the nanoantenna.

2. Theoretical background

First, we perform a theoretical analysis of optical properties of a spherical diamond nanoparticle. Despite that in our experiments nanodiamonds have complex shapes, such simplified model can provide a good qualitative results. This was demonstrated recently by comparing the theoretical and measured scattering spectra of nanodiamonds with the diameters of 250-450 nm [3].

It is known, that spherical dielectric particles support numerous resonant modes, whose frequencies and mode profiles can be found analytically. The type of resonance depends on the particle size, which in the case of diamond particle should be larger than $R \approx 100$ nm, when the first magnetic dipole resonance appears [3]. Placing the quantum emitter with luminescence frequency that matches the mode resonance frequency near the maximum of the field distribution of the mode can substantially enhance its emission rate, which is known as Purcell effect. Since at resonances the field is enhanced inside dielectric particles, we may expect the enhancement of luminescence rate for the NV-centers embedded inside the diamond particles. The calculation of emission rate can be performed with classical approach developed in Refs. [10, 11]. One can show that in small subwavelength particles (with the size $\lesssim 150$ nm in the case of diamond) the emission rate is substantially suppressed due to low local density of optical states [11], while multipole resonances with high quality factors can significantly enhance the emission rate in larger spheres [8]. In fact, due to a non-spherical shape of real diamond particles, significant roughness and possible contamination of the surface, the quality factors of these resonances and consequently Purcell factor can be reduced. This is confirmed by the absence of high-Q resonances in the experimentally measured scattering spectra. The calculations also predict that the values of the Purcell factor in different locations and for different dipole orientations can differ by several times. To take into account presence of a large number of randomly oriented emitters inside the diamond particle, we average the Purcell factor over all possible orientations of the emitters and the volume of the particle (assuming the uniform distribution of the NV-centers inside the particle), i.e. we calculate the ratio of the average radiative decay rate of the emitter inside the particle to the corresponding rate in the bulk diamond. Furthermore, the luminescence spectrum of NV-centers at room temperature spans over a broad spectral range from about 600 nm to 800 nm due to phonon sideband transitions. Strong spectral variation of the Purcell factor, caused by the resonant nature of nanoparticles, implies the same variation for the emission power. In order to estimate the possible values of the total decay rate we also normalize calculated Purcell factor to the typical luminescence spectrum of NV-centers in spectral range 620-770 nm measured in our experiments. Overall the calculation formula for

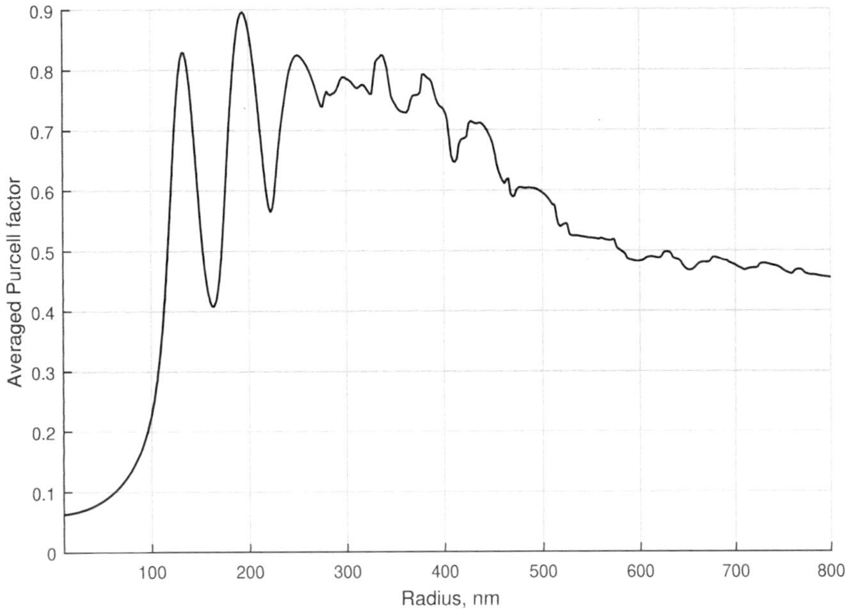

Figure 1. Purcell factor averaged over the volume of the spherical dielectric particle with permittivity $\varepsilon = 5.76 + 0.01i$ and over spectral range 620–770 nm as a function of its radius.

averaged Purcell factor $\overline{F_p}$ is the following:

$$\overline{F_p} = \frac{\int\limits_{\nu_{min}}^{\nu_{max}} \iiint\limits_{V} F_p(\mathbf{r},\nu)\, dV P_0(\nu)\, d\nu}{V \int\limits_{\nu_{min}}^{\nu_{max}} P_0(\nu)\, d\nu},$$

here $F_p(\mathbf{r},\nu)$ is the Purcell factor that depends on the position of the emitter and the emission frequency, V is the volume of the particle, $P_0(\nu)$ is the typical luminescence intensity spectrum of NV-center. Since in the measured scattering spectra of the studied particles we observed only low-Q resonances, we do not expect high values of the Purcell factor, that are predicted for lossless spherical particles with high-Q resonances. Therefore, in our calculations we introduce small material losses and describe our nanoparticles with constant dielectric permittivity ($\varepsilon = 5.76 + 0.01i$). Note, that losses were introduced only to avoid the appearance of the high-Q resonances, and therefore additional nonradiative decay rate caused by lossy medium was not taken into account. The resulting average Purcell factor is shown in Fig. 1. From these calculations, we can expect that the averaged lifetime of NV-centers in a resonant diamond particle can be reduced by $\gtrsim 3$ times as compared to nonresonant ones.

3. Experimental results

Diamonds particles were made by miling the diamond film fabricated by plasma-enhanced chemical vapor deposition (PECVD) method on a glass substrate [12, 13]. PECVD process was used with a gas source of methane (CH_4). NV-centers formation was achieved due to nitrogen implantation during the PECVD growth. Images of nanodiamonds were obtained using scanning electron microscope (SEM); an example of the nanodiamond is shown in the inset in Fig. 2(a).

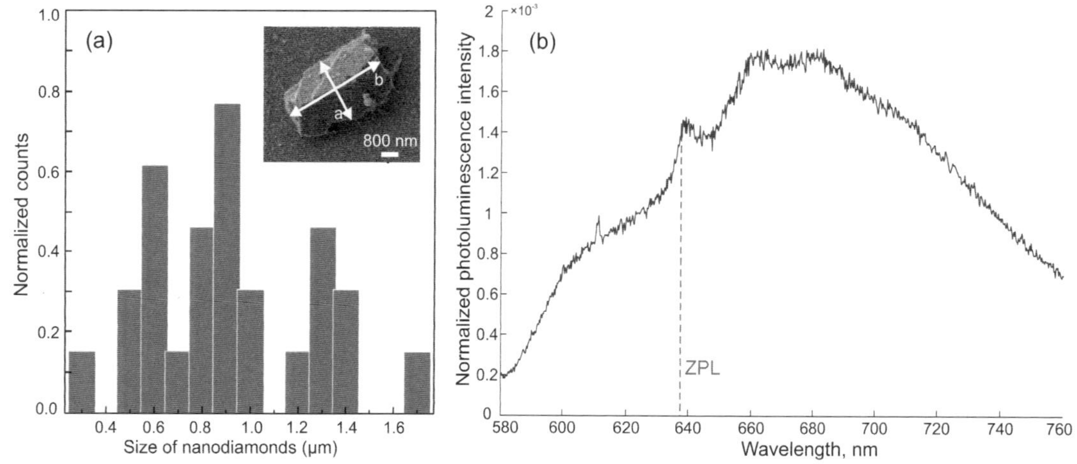

Figure 2. (a) Size distribution histogram (statistics on ∼100 nanodiamonds) measured by SEM; inset — SEM image of a nanodiamond. (b) Typical luminescence spectrum of a diamond particle with NV-centers.

The fabricated nanodiamonds had irregular shapes with rough surfaces, and we measured the characteristic sizes as shown in the insert in Fig. 2. Statistical distribution of the nanodiamond sizes is shown in Fig. 2(a), indicating median size of fabricated particles of 900 nm.

In order to study the phase composition of nanoparticles, we measured Raman scattering spectra for individual particles and applied a Gaussian fit to distinguish the characteristic Raman peaks. The analysis of nanodiamonds revealed that there were two types of nanoparticles: crystalline rough-surface and nearly spherical nanodiamonds. The Raman spectra of rough-surfaced nanodiamonds have a distinct line at 1332 cm^{-1} [14], which indicates a strong Raman scattering at the sp_3 vibrational resonance (C-C pair) of diamond lattice and proves the presence of crystalline phase of diamond. The spherical nanodiamonds that had no zero-phonon line in photoluminescence spectrum exhibited two broad asymmetric peaks at 1332 cm^{-1} and 1590 cm^{-1} — the typical vibrational resonances of surface bounds [15]. This analysis allowed us to eliminate some particles from consideration, and focus only on several particles with the pronounced and intense Raman peak at 1332 cm^{-1}. A further study of the PL spectra of selected nanoparticles reveals that most of them (more than 40 particles) demonstrate a pronounced peak at 637 nm, indicating the presence of NV-centres in the volume of the nanodiamond particle [see Fig. 2(b)].

To characterize luminescence, NV-centers were non-resonantly excited by a 532 nm laser and the lifetime measurements of the excited state were carried out by the time-correlated single photon counting method. Due to the presence of a large number of emitters randomly oriented inside the samples we have observed a nonexponential character of the time dependence of the signal, that indicates that the lifetimes of different NV-centers vary continuously in a wide range, from nanoseconds to tens of nanoseconds. Examples of time-resolved measurements are shown in Fig. 3.

Next, we compare the average lifetime in different resonant diamond particles (44 particles were measured) with the lifetime in subwavelength nonresonant nanodiamonds, for which the data is available from literature. Analytical calculations predict the severalfold enhancement of the Purcell factor for the NV-centers in large diamond nanoparticles as compared to NV-centers in subwavelength nanodiamonds. According to the literature [16, 17, 18, 5] negatively charged NV-centers in optically small ($\lesssim$ 100 nm) nanodiamonds on a glass substrate exhibit the lifetime of at least 20-25 ns. In our experiments the averaged lifetime of emitters in large

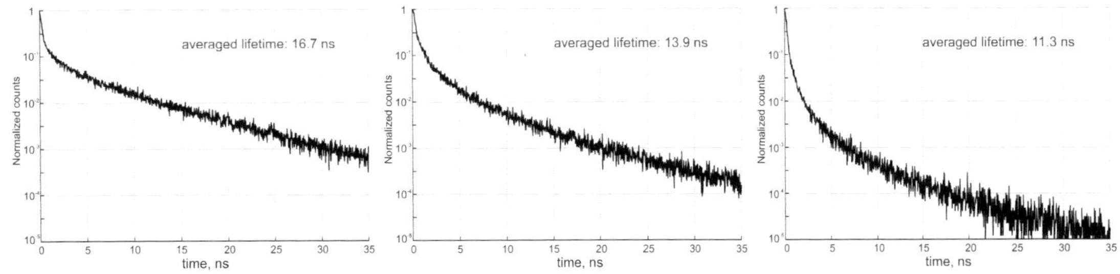

Figure 3. (a-c) Time-resolved PL measurements for three diamond particles in logarithmic scale.

resonant particles lies in the range of 10-17 ns, which is 1.5-2.5 times smaller. We also notice that the lifetime of emitters in large particles is almost unaffected by the presence of a glass substrate, while it is reduced in the case of subwavelength nanodiamonds. NV-centers lifetime in nanodiamonds placed on a porous silica substrate (with the refractive index close to 1), measured in the Ref. [17], demonstrated that the lifetime increases to more than 30 ns without a substrate. Thus, our expermiental results are in good agreement with predictions presented in Fig. 1.

4. Conclusions

We have demonstrated the possibility to control the lifetime of NV-centers via interaction with the eigenmodes of diamond-based optical nanoantennas. We have verified this experimentally for diamond particles having sizes ranging from 300×300 nm to 1.5×2 μm, supporting Mie-type resonances. We have demonstrated that the resonances of large diamond particles affect the photoluminescence properties of the NV-centers providing acceleration of their emission. We have observed a multifold decrease of the lifetime of the NV-centers in the studied diamond particles, as compared to subwavelength nanodiamonds, being in a good agreement with theoretical calculations for the average Purcell factor in the case of multiple NV-centers within a nanoparticle.

Acknowledgments

This work was supported by Russian Science Foundation, grant no. 17-72-10280.

References

[1] Kuznetsov A I, Miroshnichenko A E, Brongersma M L, Kivshar Y S and Luk'yanchuk B 2016 *Science* **354** art. no. aag2472
[2] Aharonovich I and Neu E 2014 *Adv. Opt. Mater.* **2** 911–928
[3] Shilkin D A, Shcherbakov M R, Lyubin E V, Katamadze K G, Kudryavtsev O S, Sedov V S, Vlasov I I and Fedyanin A A 2017 *ACS Photonics* **4** 1153–1158
[4] Pelton M 2015 *Nat. Photonics* **9** 427
[5] Aharonovich I, Englund D and Toth M 2016 *Nat. Photonics* **10** 631–641
[6] Shalaginov M Y, Vorobyov V V, Liu J, Ferrera M, Akimov A V, Lagutchev A, Smolyaninov A N, Klimov V V, Irudayaraj J and Kildishev A V 2015 *Laser & Photonics Reviews* **9** 120–127
[7] Faraon A, Santori C, Huang Z, Acosta V M and Beausoleil R G 2012 *Phys. Rev. Letters* **109** 033604
[8] Zambrana-Puyalto X and Bonod N 2015 *Phys. Rev. B* **91** 195422
[9] Hong H G, Lee S B, Heo M S, Park S E and Kwon T Y 2016 *Opt. Express* **24** 28815–28828
[10] Chew H 1987 *J. Chem. Phys.* **87** 1355–1360
[11] Chew H 1988 *Phys. Rev. A* **38** 3410
[12] Ivanova A, Ionin A, Khmelnitskii R, Kudryashov S, Levchenko A, Mel'nik N, Rudenko A, Saraeva I, Umanskaya S and Zayarny D 2017 *Laser Phys. Lett.* **14** 065902

[13] Pham M, Pham D, Do T, Ivanova A, Ionin A, Kudryashov S, Levchenko A, Nguyen L, Nguyen T, Rudenko A and Saraeva I 2017 *Commun. in Phys.* **27** 37
[14] Solin S and Ramdas A 1970 *Phys. Rev. B* **1** 1687
[15] Mochalin V N, Shenderova O, Ho D and Gogotsi Y 2012 *Nat. Nanotechnol.* **7** 11–23
[16] Beveratos A, Brouri R, Gacoin T, Poizat J P and Grangier P 2001 *Phys. Rev. A* **64** 061802
[17] Inam F A, Grogan M D, Rollings M, Gaebel T, Say J M, Bradac C, Birks T A, Wadsworth W J, Castelletto S and Rabeau J R 2013 *ACS Nano* **7** 3833–3843
[18] Storteboom J, Dolan P, Castelletto S, Li X and Gu M 2015 *Opt. Express* **23** 11327–11333

IOP Conf. Series: Journal of Physics: Conf. Series **961** (2017) 012008 doi:10.1088/1742-6596/961/1/012008

Controlling On-chip Optical Radiation with All-Dielectric Antennas: Reconfigurable Interconnects and Lab-on-a-chip Devices

S Lechago, C García-Meca, A Griol and J Martí

Nanophotonics Technology Center, Universitat Politècnica de València, Camino de Vera s/n, 46022 Valencia, Spain.

E-mail: serlecbu@ntc.upv.es

Abstract. Aimed to improve the flexibility of optical network-on-a-chip topologies, unguided optical interconnects using plasmonic nanoantennas or dielectric phased arrays have been proposed. However, the bulky footprints of the latter, and both the low directivity figures and high losses of the former, together with complicated excitation schemes, limit their use for on-chip optical interconnects. Here, we introduce a novel concept of on-chip CMOS-compatible wireless optical system based on the use of smartly-engineered broadband easily-fed antennas, which not only overcomes the aforementioned drawbacks but also opens a wide range of applications in several fields. To illustrate its potential, several unprecedented on-chip wireless applications are outlined and experimentally demonstrated. This includes the verification of broadband highly-directive wireless data transmission at speeds as high as 160 $Gbit{\cdot}s^{-1}$ over mm-scale links, the realization of fully-reconfigurable wireless beam steering device and the validation of an ultra-compact integrated contactless microflow cytometer.

The ceaseless progress in the field of CMOS ultra-large system integration is continuously requiring more energy-efficient and flexible on-chip interconnection networks [1], manufactured at ever tinier scales with pinpoint accuracy. It is well-known that on-chip guided photonic systems can deliver ultra-high-speed low-power solutions for both multi-core interconnections and traffic routing [2].

However, this kind of waveguide-based photonic architectures is not able to provide the necessary flexibility to relax the strict interconnecting design rules of the optical network-on-chip and seamlessly pave the way for a much wider range of on-chip applications. Such a limitation stems precisely from the use of waveguides, either plasmonic or dielectric, which confer these networks a fixed static physical topology. Along this line, trying to improve this lack of flexibility, photonic unguided links at the nanoscale have been proposed using plasmonic nanoantennas, as mimicking radiofrequency links [3]. In fact, plasmonic nanoantennas [4,5] have recently emerged as a key piece in nanophotonic circuit engineering due to their capacity to concentrate optical energy at very high frequencies and to prove large spontaneous emission enhancement [6]. Unfortunately, on the one hand, plasmonic nanoantennas exhibit poor directivity and high losses as a result of their strong field confinement and metallic absorption at optical frequencies [7]. Additionally, the excitation of these nanoantennas usually requires complex feeding elements such as quantum dots or nanodipoles [8], which experimentally represents an additional technological challenge. Likewise, we can also find studies based on the use of dielectric antenna arrays [9,10], implemented on vertical grating couplers that attempt to offer a solution for optical data interconnects. From a general perspective, while these configurations allow higher directivity figures, they usually lead to less compact devices, more focused on off-chip data transmission.

The current shortages associated with these implementations reveal that a high-performance new model of wireless optical interconnecting architecture is required to bring efficient solutions in a wide variety of applications. Moreover, such architecture could not only fulfil the current photonic waveguide-based systems demands but also open ground-breaking on-chip photonic applications in several fields. In this work, building on a new concept of smartly-engineered silicon antennas we show the feasibility of the aforesaid high-performance on-chip flexible wireless optical architecture by experimentally demonstrating the first wireless long-reach high-speed on-chip data transmission, a fully-reconfigurable wireless optical system and a contactless photonic microparticle sensor.

The first step for achieving the desired architecture is to develop a kind of antenna exhibiting low loss and low reflections, and whose directivity can be tuned to take a wide range of values able to accommodate different applications, while always keeping a reduced size. For instance, for high-speed data transmission wireless optical links, the directivity must be as high as possible (well beyond the figures reported so far) to increase the optical link reach up to the mm-scale. Recent studies show that nanoantennas based on dielectric nanoparticles exhibit lower losses and a higher directivity than metallic structures [11], although still too low for the above-mentioned applications. Moreover, these structures present important fabrication challenges. Following this thread, we studied the possibilities of open-end dielectric waveguides as potential radiating elements. Using Huygens' principle [12], we modelled this kind of structures as aperture antennas, in the same way that microwave metallic waveguides are analyzed. This simple approach links directly the radiation pattern of the waveguide with the propagating mode profile, providing simple design rules that surprisingly allow us to tailor the antenna directivity from values as low as 5 to figures higher than 100 [13]. In addition, the weakly confined modes supported by these structures favor impedance matching, thus minimizing reflections. Finally, the use of dielectric materials guarantees the desired low losses. With this reasoning in mind, we designed several structures based on well-known silicon photonic waveguides [14], namely, strip and slot configurations, ensuring a simple fabrication process using conventional dielectric elements and avoiding other complex feeding structures, usually employed with plasmonic nanoantennas.

We focused primarily on the strip option, since in this case the antenna directivity can be easily tuned by gradually reducing its width through an inverted-taper configuration, which allows a reflectionless transition from a standard 450-nm-wide strip to the radiating end. For the sake of fabrication simplicity and to avoid long taper structures we fixed a minimum taper tip width of 150 nm, which allowed us to attain directivities higher than 50 at near-infrared wavelengths. Moreover, we verified that by using additional elements (directors) [13,15], it is possible to further improve the directivity while keeping a reduced antenna size, see figure 1(a). After an optimization process, we obtained a director configuration consisting of two symmetric silicon strips. The proper combination of a single antenna and the designed directors enhanced the directivity to figures as high as 114. In order to numerically calculate the directivity of the antennas, the fundamental TE mode of a standard 450-nm-width 220-nm-height silicon strip waveguide is excited. This waveguide is directly butt-coupled to the input facet of the antenna, see figure 1(b), being the responsible of feeding the structure, which in turn will radiate under far-field conditions. So as to excite the aforementioned TE mode, standard ports were set up at the input facets to simulate infinitely long waveguides. The rest of the computational domain was terminated with perfectly matched layers (open boundaries) to avoid undesirable reflections.

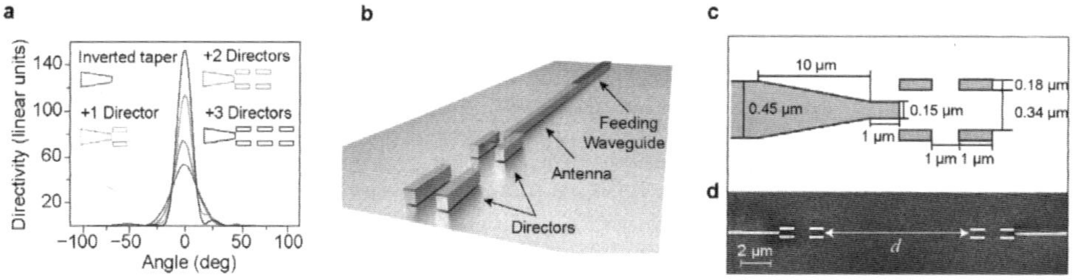

Figure 1. (a) Directivity of the designed antenna as a function of the number of directors. (b) Artwork of the proposed final silicon antenna design. (c) Schematic top view of the final antenna configuration and directors. (d) SEM image of a wireless point-to-point antenna link with d = 10 μm. Calculations were performed by using the CST Microwave Studio Suite.

Note that recent dielectric antennas [11] reach directivity values up to 12, while typical plasmonic nanoantennas exhibit lower values [16]. The final configuration of the antenna including the directors is displayed in figure 1(b). The optimized dimensions of the structures as well as a SEM image of a fabricated link are depicted in figures 1(c) and 1(d), respectively. Note that, as these structures do not rely on resonant phenomena a large bandwidth is also guaranteed. Additionally, these dielectric structures display almost perfect impedance matching in a wide spectral region, as well as a total efficiency close to 90%. Thus, unlike in typical plasmonic works, auxiliary resonant matching elements are not needed, avoiding the bandwidth limitation they introduce. Let us now see how this unprecedented low-loss, broadband, and ultra-directive configuration (not so far achieved with a single optical antenna) can provide novel efficient solutions in a variety of situations.

As a first example, using the previously shown final antenna configuration including two directors, we experimentally achieved, for the first time, a long-reach high-speed wireless data transmission through mm-scale on-chip links consisting of two identical antennas. For this purpose, several samples with different gap distances d (0.1, 0.2, 0.4, and 0.8 mm) were fabricated on a silicon-on-insulator platform. We measured the link power efficiency (see figure 2(a)) as a function of d by injecting light into standard 450-nm-wide strip waveguides (connected to the antennas) using conventional grating couplers and tapers. The experimental antenna gain was also retrieved (see figure 2(b)) and found to be in good agreement with our theoretical estimation. Digital data transmission over the 0.4-mm link was demonstrated by sending 40 Gbit·s^{-1} pseudo-random bit sequences at 38 different ITU channels evenly spaced by 0.8 nm, ranging from 1534.25 to 1563.86 nm. We measured bit error rates (BER) below 10^{-9} (sufficiently low to rule out forward-error correction schemes often related to an increase in chip complexity) in all cases, verifying the above-mentioned broadband features. This performance shows the feasibility of attaining terabit-per-second data streaming over mm-scale distances by using wavelength multiplexing.

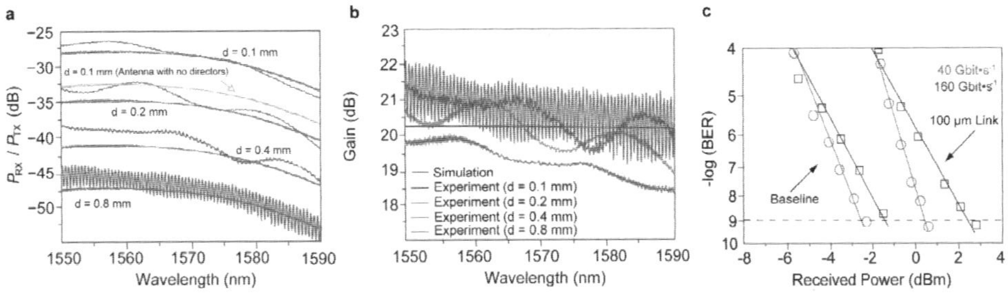

Figure 2. (a) Experimental (blue) and theoretical (red) power efficiency for different distances *d*. Theoretical results were retrieved by using the Friis transmission equation: PRX = PTX·GTX·GRX·(λ/(4πd)), where PRX and PTX are the received and transmitted power of the antennas, respectively. GTX and GRX stand for the gain of the transmitting and receiving antennas, λ is the working wavelength and d is the link distance between the antennas. For comparison purposes, the simulated efficiency corresponding to a bare tapered antenna with no directors is also included for the 100-μm link. (b) Simulated and experimental (retrieved from the measurements) antenna gain. (c) BER measured with an error analyzer and evaluated as a function of the decreasing optical power received by the photodetector for a baseline configuration where the wireless link was replaced with a silicon straight waveguide (left) and for a 100-μm wireless link (right). The 160 Gbit·s^{-1} BER corresponds to the demodulation of the wavelength corresponding to ITU channel 34 when transmitting channels 33 to 36. The crosstalk produced by surrounding channels (33, 35 and 36) introduces low penalties.

To illustrate this possibility, we additionally transmitted 160 Gbit·s^{-1} data streams consisting of four de-correlated 40-Gbit·s^{-1} sequences at wavelengths ranging from 1548.52 nm up to 1550.92 nm (0.8 nm evenly distributed), obtaining the same performance in terms of BER, see figure 2(c). Remarkably, the ultra-directive behavior here shown could permit the implementation of unguided crossings in many different configurations without perturbing the beam propagation or adding any undesired crosstalk (and even leads us to envisage the possibility of achieving efficient inter-core on-chip wireless interconnects), thus providing a much more flexible optical interconnect architecture.

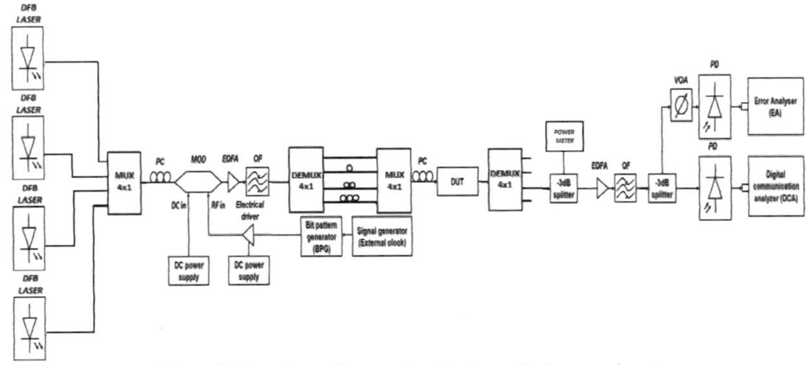

Figure 3. Experimental set-up for data transmission experiments

Figure 3 shows a scheme of the set-up employed for the data transmission experiment. Measurements were carried out using a commercial $LiNiBO_3$ modulator fed with a 40 Gbit s^{-1} non-return to zero (NRZ) pseudo-random bit sequence (PRBS 2^{31}-1) delivered by a bit pattern generator (SHF BPG 44E). The electrical signal was amplified through a broad bandwidth driver amplifier (DC to 40 GHz) to achieve a voltage swing of ~5 V_{pp} and combined to a 1.7 V DC offset to ensure the modulator is biased at quadrature. The input Gbit·s^{-1} modulated signal was then coupled in and out the silicon chip (DUT) via grating couplers, with typical insertion losses below 9 dB at λ = 1550 nm. In addition, silicon tapers were employed to couple

IOP Conf. Series: Journal of Physics: Conf. Series **961** (2017) 012008 doi:10.1088/1742-6596/961/1/012008

light from the grating to standard 450 nm-wide strip waveguides and vice versa. The signal transmitted through the link was photodetected by a 40 GHz Digital Communication Analyser (Infiniium DCA-J 86100C)

The ever more indispensable requirement for reconfigurable on-chip networks motivated a second application: a dynamically-tunable antenna beam-steering system. To this end, we benefited from a distinctive feature of the proposed architecture, i.e., the excitation scheme based on silicon waveguides. The fact that the refractive index of these structures can be modified dynamically via different effects (e.g., thermo-optic, electro-optic) permits us to tune independently the phase α (and ultimately the amplitude via interferometry) of each antenna's input signal. Consequently, the use of antenna arrays in combination with any of these effects makes it possible to mold the in-plane radiation diagram in almost any desired way. As an example, we exploited silicon thermo-optic effect [17] through the use of metallic nano-heaters, each terminated by two pads and placed atop the feeding waveguides (separated by a 1-μm SiO_2 layer from the substrate), to control the input phase (α) of each antenna [13]. This is achieved by modifying the temperature of each waveguide through the voltage applied across a corresponding nano-heater. With this simple technique, we experimentally demonstrated a reconfigurable 30-degree beam-steering system over 100-μm links (λ = 1550 nm). Figures 4(a) and 4(b) show SEM images of the transmitting and receiving beam-steering subsystems. The device constitutes a 1-to-3-port switch and, contrary to existing guided solutions [18], is completely scalable thanks to its wireless nature, i.e., more inputs/outputs can be added without additional insertion losses [13]. In addition, we estimated the device energy consumption and speed, finding a switching power of 8.3 mW for a single heater (three times this value for the complete beam steering device) and a switching time of 5.2 μs. Note that both values are in line with current state of the art. Moreover, the sharp beams radiated by our ultra-directive antennas yield a very good crosstalk figure (see figure 4(c)). That is, when a beam is steered towards a determined antenna, the power drastically decreases in the non-receiving ones. It is worth to point out that this feature could not be achieved with low-directivity antennas, invalidating the steering device for many applications such as data switching.

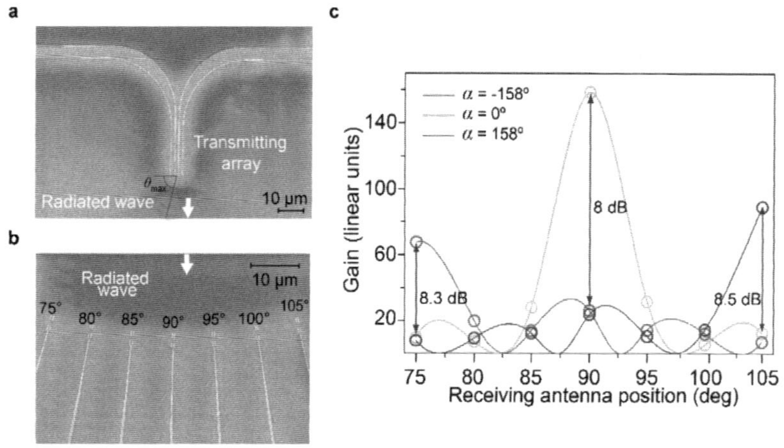

Figure 4. (a) SEM image of the waveguides used to feed the 4-antenna emitting array. (b) SEM image of the receiving subsystem consisting of seven antennas placed at different angles. The antennas at 75°, 90° and 105° are taken as the device output ports. The values of α that allow us to steer the beam to each of these ports were estimated to be − 158°, 0° and 158°, respectively. (c) Total array gain associated with each receiving antenna (d=100 μm, λ=1550 nm) for the three values of α. Circles indicate measured values at each antenna position. The crosstalk at the output ports is highlighted, with values always better than 8 dB.

As a final example, inspired by the technique employed in flow cytometry [19], we designed and fabricated an on-chip hydrodynamic parametric analysis system in which different kinds of microparticles can flow through a microfluidic channel and be simultaneously illuminated by a narrow beam. The main goal was to classify the flowing particles according to their unique time-dependent scattered-field signatures. To this end, we incorporated an optical subsystem to the chip consisting of a highly-directive emitting antenna (inverted taper + 2 directors) and two identical receiving antennas connected to standard strip waveguides, see figure 5(a). One of the receiving antennas collects the power at 90° and is used for alignment/calibration purposes. The second receiving antenna measures the scattered field at 60°, which enables us to classify the particles. On the other hand, a microfluidic channel was built to allow the particles to flow in between the antennas. To test the device, we used polystyrene microspheres (a standard

IOP Conf. Series: Journal of Physics: Conf. Series **961** (2017) 012008 doi:10.1088/1742-6596/961/1/012008

benchmark for calibrating flow cytometers) with two different diameters (1 and 2 μm) in an aqueous solution with a concentration of 1:100 (2% solids). Capillary forces ensure that the solution fills the channel and the microspheres flow between the antennas when casting a solution drop into the fluidic system. To learn the characteristic scattered-field signature of each kind of microsphere, we prepared two different solutions including either 1-μm spheres or 2-μm spheres alone. Both signatures appear in the form of a dip, of approximately 2 dB in the case of the large microspheres and of 0.25 dB in the case of the small ones, see figures 5(b) and 5(c). The signatures are exclusive, as no 2 (0.25) dB dips were observed when flowing the 1-μm (2-μm) spheres. Finally, an equally-mixed solution containing both kinds of particles was flowed, see figure 5(d). In this case, we were able not only to detect when a particle goes through the optical link (due to the presence of a dip), but also to easily classify such a particle, i.e., identify its size, depending on the magnitude of the measured dip. It is worth mentioning that, since the optical system is not in physical contact with the target, it can be reused an unlimited number of times. Outstandingly, the directive nature of the illuminating antenna implies that very small particles can be resolved, providing a promising solution for sub-micron particle detection. The proposed device constitutes a low-cost chip-integrated type of flow cytometer [20], much more compact than current versions, since no bulky elements such as optical fibers or lenses are required, eliminating potential losses and failure of discrete component interfaces and simplifying the packaging and the automation of the manufacturing process. Finally, the high density of detectors allowed by the small size of the antennas permits a finer scattered-field sampling angular resolution, resulting in a more accurate classification system if necessary.

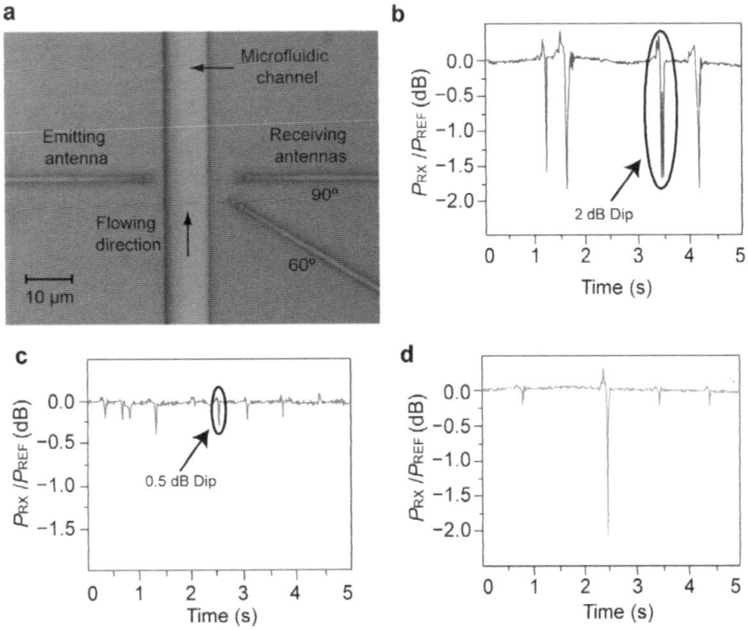

Figure 5. (a) Optical microscope image of the fabricated device. (b–d) Power time-dependent efficiency measured at 60° for 1-μm (b), 2-μm (c), and a mixed solution of polystyrene microspheres (d).

In conclusion, we have experimentally demonstrated a new concept of flexible and reconfigurable wireless optical system built upon a disruptive kind of high-performance silicon optical antennas. This approach provides a CMOS-compatible on-chip platform supporting a wide range of multi-purpose ultra-compact applications hardly manageable via traditional waveguide-based architectures or plasmonic-based implementations, hence opening a new era for the future nanophotonic circuitries and technologies. These applications include flexible wireless high-speed data transmission links and dynamically-reconfigurable optical network-on-chip topologies, with functionalities ranging from antenna beam steering or beam focusing to 2D-selective optical detection or 2D space mapping, as well as integrated contactless sensing systems for realizing lab-on-chip devices.

Fabrication of the samples

The antenna links were fabricated on standard SOI samples from SOITEC wafers with a top silicon layer thickness of 220 nm (resistivity $\rho \sim$1-10Ω·cm^{-1}, with a lightly p-type background doping of $\sim 10^{15}$ cm^{-3}) and a buried oxide layer thickness of 2 μm. The fabrication is based on an electron-beam direct-writing process performed on a coated 100 nm hydrogen silsesquioxane (HSQ) resist film. The mentioned electron-beam exposure, performed with a Raith150 tool, was optimised in order to reach the required dimensions employing an acceleration voltage of 30 KeV and an aperture size of 30μm. After developing the HSQ resist using tetramethylammonium hydroxide, the resist patterns were transferred into the SOI samples employing an optimised Inductively Coupled Plasma-Reactive Ion Etching process with fluoride gases. Finally, a two-micron-thickness silicon dioxide uppercladding was deposited on the SOI sample by using a Plasma Enhanced Chemical Vapour Deposition system from Applied Materials. For the fabrication of the fluidic channel, a Cr layer of 35 nm was first deposited on the SOI sample by using electron beam metal evaporation. A direct writing electron beam exposure of the channel was then performed on a layer of 100 nm of PMMA 950K positive resist. After resist developing, Cr was removed from the channel area using a wet Cr etchant process based on CR-7 ($HClO_4 + C(NH_4)_2(NO_3)_6 + H_2O$) . Afterwards, an ICP-RIE process was carried out to open the channel through the SiO_2. Finally, the sample was cleaned in order to remove the organic residues by a mixture of H_2SO_4 and H_2O_2 (3:1) during 20 minutes and then washed by deionized water (DIW). This cleaning procedure was also used to regenerate the device after sensing experiments. To prepare the employed PDMS thin substrates (Sylgard 184 Silicone Elastomer Down Corning), the PDMS mixture (10:1) was spin-coated on glass cover slides and cured at 60 °C during 1 hour to obtain a thin (150 μm) PDMS layer on the glass slide. After the curing process, the PDMS was peeled off from the glass slide and washed in absolute ethanol to clean the surface. Finally the PDMS substrate was positioned and aligned on top of the fluidic channel.

References

[1] Haurylau M, Nelson N A and Friedman EG 2006 *IEEE J. Quantum Electron.* **12** 1699.
[2] Chan J, Hendry G, Biberman A and Bergman K 2010 *J. Lightw. Technol.* **28** 1305.
[3] Alù A and Engheta N 2010 *Phys. Rev. Lett.* **104**, 213902.
[4] Novotny L and van Hulst N F 2011 *Nature Photon.* **5** 83.
[5] Fischer H and Martin O J F 2008 *Opt. Express* **16** 9144.
[6] Liaw J W 2008 *IEEE J. Quantum Electron.* **14** 1441.
[7] Bharadwaj P, Deutsch B and Novotny L 2009 *Adv. Opt. Photon.* **1** 438.
[8] Curto AG, Volpe G, Taminiau T H, Kreuzer M P, Quidant R and van Hulst N F 2010 *Science* **329** 930.
[9] Van Acoleyen K, Rogier H. and Baets R. 2010 *Opt. Express* **23** 13655.
[10] Sun J, Timordogan E, Yaacobi A, Hosseini E S and Watts M R 2013 *Nature* **493** 195.
[11] Krasnok A E, Miroshnichenko A E, Belov P A and Kivshar Y S 2012 *Opt. Express* **20** 20599.
[12] Balanis C 1982 *Antenna Theory: Analysis and Design.* (Wiley, New York).
[13] García-Meca C, Lechago S, Brimont A, Griol A, Mas S, Sánchez L, Bellieres L, Losilla N S and Martí J 2017 *Light Sci. Appl.* **6** e17053.
[14] Cárdenas J, Poitras C B, Robinson J T, Preston K, Chen L and Lipson M 2009 *Opt. Express.* **17** 4752.
[15] Kosako T, Kadoya Y and Hofmann H F 2010 *Nature Photon.* **4** 312.
[16] Dregely D, Lindfors K, Lippitz M, Engheta N, Totzeck M and Giessen H 2014 *Nature Commun.* **5** 4354.
[17] Yun S S and Lee J H 2003 *J. Micromech. Microeng.* **13** 721.
[18] Lee B G *et al.* 2014 *J. Lightw. Technol.* **32** 743.
[19] Robinson J P, Roederer M 2015 *Science* **350** 739.
[20] Zhang YS, Watts B R, Guo T Y, Zhang Z Y, Xu C Q and Fang Q 2016 *Micromachines* **7** 70.

IOP Conf. Series: Journal of Physics: Conf. Series **961** (2017) 012009 doi:10.1088/1742-6596/961/1/012009

Interface defects and silicon impurities in GaAs/$Al_{0.3}Ga_{0.7}As$ heterostructures

Y V Lebiadok[1], **A A Shalayeva**[1]

[1] State Scientific and Production Association "Optics, Optoelectronics and Laser Technology", Belarus, Minsk, Niezaliežnasci Ave., 68

E-mail: y.lebiadok@ifanbel.bas-net.by

Abstract. The formation of complex defects (gallium, arsenic and aluminium vacancies with corresponding interstitial atoms) present at the GaAs/$Al_{0.3}Ga_{0.7}As$ heterointerface as well as passivation of the defects by silicon impurities are discussed in the report. We used the density functional theory calculations, with the hybrid functionals B3LYP with Hay-Wadt effective core potentials for all the heavy atoms, in combination with Hay-Wadt valence basis. We present the energy characteristics (formation energy) and geometry of the defects (spatial distribution of the atoms and its charge near the interface defect).

1. Introduction

Infrared photodetectors based on GaAs/AlGaAs quantum wells (absorption wavelength rage of 8-10 μm) are widely used for development of military thermal vision systems. One of the main requirements for the systems is reliability which depends on the presence and characteristics of defects in the GaAs/AlGaAs heterostructure. Point defects, as well as the doping process, are of great interest due to the influence on the structure of quantum wells of GaAs/AlGaAs, as well as changes in the processes of charge transfer and energy levels. With the change in the structure of quantum wells, the parameters of infrared photodetectors change [1].

The main alloying impurities in GaAs/AlGaAs are silicon and carbon. Doping of AlGaAs barriers by Si atoms not only leads to a change in the structural, but also optical properties in the infrared spectral range. In the absorption spectra of GaAs/AlGaAs heterostructure additional bands arise, associated with the appearance of complex defects involving silicon. On the other hand doping of AlGaAs by Si can lead to passivation of structural defects (vacancies) which leads to improving reliability of the devices based on GaAs/AlGaAs heterostructure.

This paper presents the energy characteristics, as well as the spatial distribution of the defect complexes in the GaAs/AlGaAs heterostructure of quantum well infrared photodetectors (QWIP).

2. Methods of Calculation

To determine the energy and structural characteristics of point defects, the cluster approximation method was used [2]. For the GaAs/AlGaAs cluster, the GaAs crystal lattice (zinc blende) was chosen, with lattice constant at 300 K [3]. The model clusters were formed by "cutting" the bounded fragment

IOP Conf. Series: Journal of Physics: Conf. Series **961** (2017) 012009 doi:10.1088/1742-6596/961/1/012009

from the GaAs crystal volume [4]. AlGaAs was modelled by replacing the gallium atoms by aluminium atoms (in a percentage ratio of about 30%), taking into account the growth of the real crystal (in our case, the growth plane of GaAs is (100)) [5]. The cluster GaAs/AlGaAs consisted of 72 arsenic atoms, 62 gallium atoms and 10 aluminium atoms (total 144 atoms). The bonds on the cluster boundary were closed by hydrogen atoms, with the aim of eliminating distortion of the structure and improving the values of convergence [4].

Geometry optimization of the cluster was carried out on the base of the density functional theory (with the hybrid functional B3LYP1 and the relativistic Hay-Wadt potentials). Atomic orbitals were given by the basic kit Hay-Wadt [6]. Calculations were made in the software package FireFly [7].

Complexes of defects were created at the interface of the well/barrier $GaAs/Al_{0.3}Ga_{0.7}As$ by transferring the atom from the crystal lattice site (interstitial atoms), as well as removing the atom from the lattice site and adding an interstitial silicon atom (for the vacancy – interstitial silicon atom complexes). The full geometry optimization with respect to total energy minimization was performed after construction of the model cluster (gradient convergence tolerance for the calculation was taken as 0.00001 Hartree/Bohr (this value is optimal for the clusters based on A_3B_5 compounds [8, 9]).

The formation energies of the defect complexes were determined by the difference in the total energy of the defect-free cluster relative to the total energy of the optimized defective cluster.

3. Results and discussion

In the case of the gallium atom moved to the interstitial position within the first coordination sphere, the cluster geometry optimization leads to annihilation of the vacancy with an interstitial gallium atom [10]. This means that the defect is not stable. Thus, to determine the stable configuration of defects it is appropriate to consider the complex of defects, namely, vacancy + the interstitial atom in second coordination sphere, for example, the complex of the gallium vacancy V_{Ga} and its interstitial atom Ga_i in the crystallographic plane (100), as shown on figure 1.

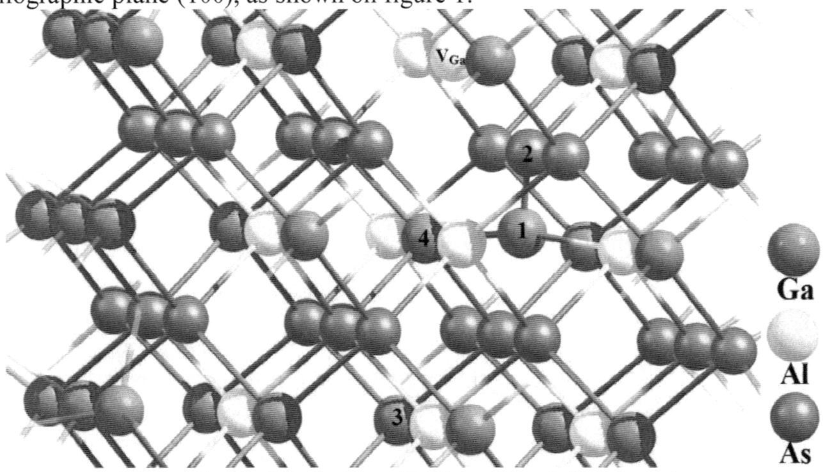

Figure 1. Model cluster with $Ga_i – V_{Ga}$ defect complex.

For above motioned complex defect the arsenic atom (2) moves to the V_{Ga} position, the gallium atom (1) replaces the arsenic atom (2) during the geometry optimization procedure. As a result, two anti-structural defects of As_{Ga} and Ga_{As} are formed [11]. The gallium atom (3) moves to the interstitial position, Ga_i as a defect is preserved. The energy of formation of this complex is 4.9295 eV [12].

The interatomic distances for the initial cluster (r_c) and the geometrically optimized model (r_o) for a cluster with $Ga_i – V_{Ga}$ defect complex are shown in table 1. The distances r_o represent the spatial parameters of the defect and can be determined on the base of experimental investigations, for example, using extended X-ray absorption fine structure spectroscopy [9].

IOP Conf. Series: Journal of Physics: Conf. Series **961** (2017) 012009 doi:10.1088/1742-6596/961/1/012009

Table 1.Interatomic distances of cluster $Ga_i - V_{Ga}$ defect complex.

	2-1	1-4	4-3	4-2
r_c, Å	1.28	1.80	4.00	2.45
r_o, Å	2.64	2.57	2.60	4.15

A complex of defects "arsenic vacancy V_{As} and its interstitial atom As_i" in the crystallographic plane (100) is shown on figure 2.

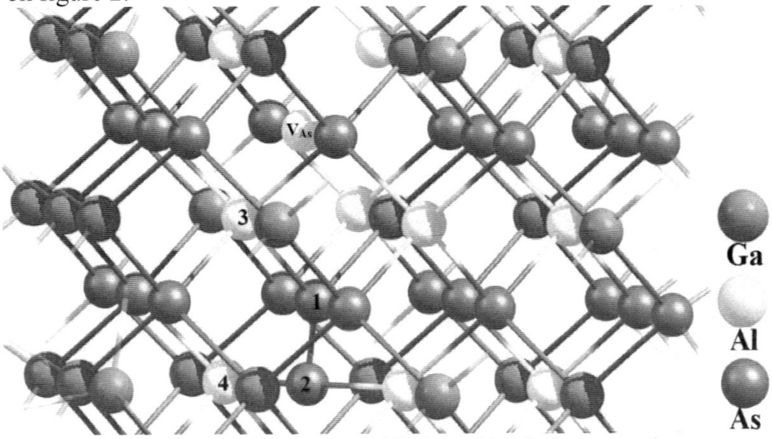

Figure 2. Model cluster with $As_i - V_{As}$ defect complex.

In the process of geometric optimization, two vacancy-interstitial arsenic (As_i-V_{As}) complexes are formed. Atom 1 moves to the interstitial position, but its place remains the vacancy of arsenic. The energy of formation of this complex is 6.034 eV [13]. The interatomic distances for the initial cluster (r_c) and the geometrically optimized model (r_o) for a cluster with $As_i - V_{As}$ defect complex are shown in table 2.

Table 2.Interatomic distances of cluster $As_i - V_{As}$ defect complex.

	2-1	2-4	1-3	3-4
r_c, Å	1.44	1.73	2.45	4.00
r_o, Å	2.49	2.54	2.57	4.70

To investigate the possibility of passivation of V_{As} and V_{Ga} by silicon impurity the complex of defects "vacancy and the interstitial silicon atom Si_i" were considered. The defects "gallium vacancy and the interstitial silicon atom Si_i" is shown in figure 3.

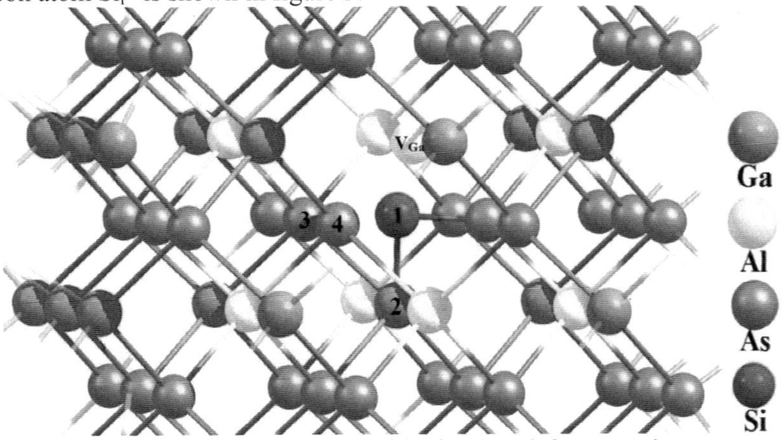

Figure 3. Model cluster with $Si_i - V_{Ga}$ defect complex.

In the process of geometric optimization, a more complex defective complex appears: Si_i and the complex V_{Ga} and Ga_i. The atom (2) moves to the interstitial position, V_{Ga} remains in its place. The charge density of cluster with $Si_i - V_{Ga}$ defect complex in the crystallographic plane (011) is shown of figure 4. The charge distribution in the vicinity of the defect can be used to calculate the defect charge and the energy levels of the defect in the $GaAs/Al_{0.3}Ga_{0.7}As$ quantum well (the energy levels determine absorption spectrum of the QWIP).

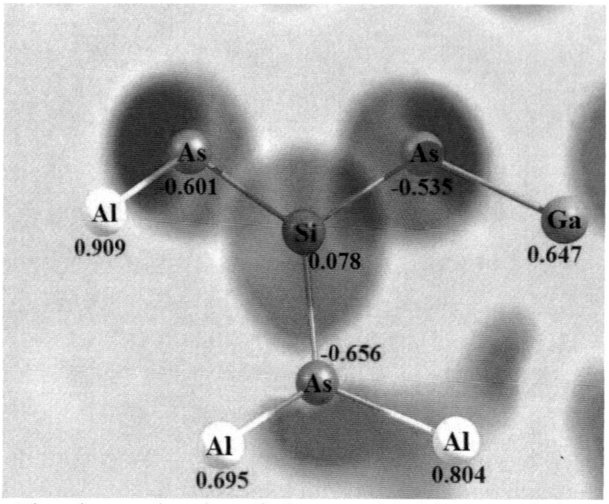

Figure 4. The charge density of cluster with $Si_i - V_{Ga}$ defect complex.

At is seen in figure 3 and 4 the charge of the silicon atom does not correspond to the lattice position of atoms (for example, arsenic and gallium), which confirms the preservation of the interstitial silicon atom as a defect. The interatomic distances for clusters with $Si_i - V_{Ga}$ defect complex are shown in table 3.

Table 3. Interatomic distances of cluster $Si_i - V_{Ga}$ defect complex.

	2-1	1-4	2-3	3-4
r_c, Å	1.79	3.54	2.45	5.65
r_o, Å	2.51	2.48	2.52	5.97

The defect complex "arsenic vacancy and the interstitial silicon atom" in the crystallographic plane (100) is shown on figure 5.

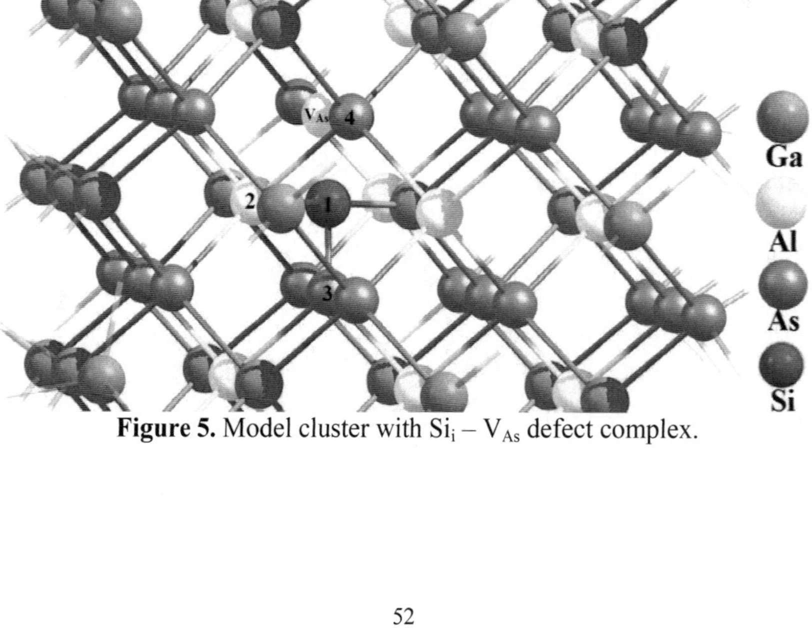

Figure 5. Model cluster with $Si_i - V_{As}$ defect complex.

For this case in the process of geometric optimization the interstitial silicon atom annihilates with an arsenic vacancy (atom 1 replaces V_{As}), resulting in the formation of an anti-structural defect of Si_{As} [14]. This means that silicon impurity passivates the arsenic vacancy. The charge density of cluster with $Si_i - V_{As}$ defect complex in the crystallographic plane (100) is shown of figure 6.

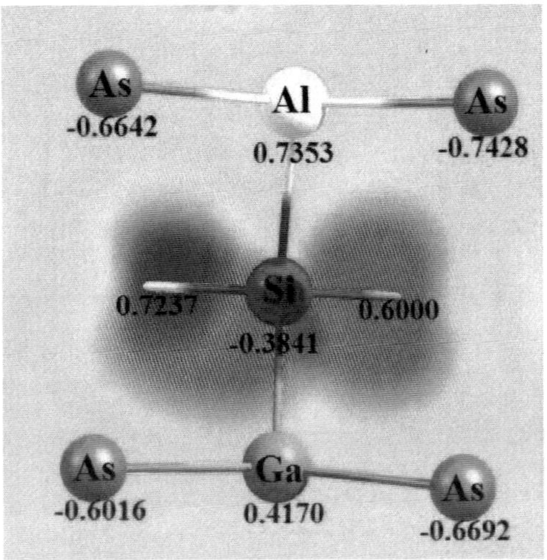

Figure 6. The charge density of cluster with $Si_i - V_{As}$ defect complex.

On the base of the distribution of the charge in the near-defect region cluster with the defect $Si_i - V_{As}$ one can conclude that the charge of the silicon atom roughly corresponds to the charge of the lattice atom of arsenic (this fact confirms the occurrence of a new Si_{As} defect in the process of geometric optimization). The interatomic distances for clusters with $Si_i - V_{As}$ complex defects are shown in table 4.

Table 4. Interatomic distances of cluster $Si_i - V_{As}$ defect complex.

	2-1	1-3	1-4	2-4
r_c, Å	2.31	1.46	3.58	2.45
r_o, Å	5.02	4.20	5.99	2.58

4. Conclusions

On the basis of the calculations carried out, it was shown that stable complex defects As_i-V_{As} and $As_{Ga} + Ga_{As}$ are formed in GaAs/$Al_{0.3}Ga_{0.7}$As heterostructure. It was ascertained that an impurity silicon atom passivates the vacancy of an arsenic atom, but does not passivates the vacancy of a gallium atom. Thus, the alloying of the barrier layer of GaAs/$Al_{0.3}Ga_{0.7}$As quantum well by silicon atoms leads to improving reliability of GaAs/AlGaAs quantum well infrared photodetector when the QWIP is used under a radiation influence conditions (for the radiation flux which causes an arsenic vacancy formation). The charge states of silicon impurities in the GaAs/$Al_{0.3}Ga_{0.7}$As heterostructure were calculated also. The data obtained can be used for simulation of the QWIP absorption spectrum.

References

[1] Sengupta D, Jandhyala V, Kim S, Fang W, Malin J, Apostolakis P, Kwong-Chi Hseih, Yia-Chung Chang, Shun Lien Chuang, Bandara S, Gunapala S, Feng M, Michielssen E and Stillman G 1998 *IEEE* **4** 4 p. 746–757
[2] Yip S 2005 *Handbook of materials modeling* (Dordrecht: Springer) p. 15
[3] Vurgaftman I, Meyer J R, Ram-Mohan L R 2001 *J. Appl. Phys.* **89** 11 p. 5815–5875
[4] Seel M. 1988 *Int. J. Quant. Chem.* **34** (s22) p. 265–274

[5] Butyagin O F, Katsavets N I, Kogan I V, Krasovitsky D M, Kulikov V B, Chalyi V P, Dudin A L, Cherednichenko O B 2012 *Technical Physics Letters* **38** 5 p. 436–438
[6] Hay P J, Wadt W R 1985 *J. Chem. Phys.* **82** p. 283–298
[7] http://classic.chem.msu.su/gran/gamess/index.html
[8] Bezyazychnaya T V, Kabanau D M, Kabanov V V, Lebiadok Y V, Ryabtsev A G, Ryabtsev G I, Zelenkovskii V M, Mehta S K 2015 *Lithuanian Journal of Physics* **55** 1 p. 10–16
[9] Lebiadok Y V, Bezyazychnaya T V, Zhuravlev K S 2016 *Proc. of SPIE* **9994** p. 99940I-1–6
[10] Wee S F, Chai M K, Homewood K P, Gillin W P 1997 *J. Appl. Phys.* **82** p. 4842–4846
[11] Tan T Y, Gösele U 1988 *Mater. Sci. Eng.* **B** 1 p. 47–65
[12] Baikov V I, Isaev E I, Korzhavyi P A, Vekilov Yu Kh, Abrikosov I A 2005 *Physics of the Solid State* **47** 10 p. 1831–1836
[13] Northrup J E, Zhang S B 1993 *Phys. Rev.* B **47** 11 p. 6791-6794
[14] Talwar D N, Vandevyver V 1989 *Phys. Rev.* B **40** 14 p. 9779-9789

IOP Conf. Series: Journal of Physics: Conf. Series **961** (2017) 012010 doi:10.1088/1742-6596/961/1/012010

Colour gamut enhancement with remote light conversion mechanism

D Koseoglu[1,2], Y S Sezer[1,2] and K Karsli[1]

[1]Department of Display Design, Vestel Electronics, 45030, Turkey
[2]Department of Electrical & Electronics Engineering, Ozyegin University, 34794, Turkey

Abstract

The backlight unit spectrum of liquid crystal displays (LCD) directly affects the colour gamut. With the invention of GaN based blue light emitting diodes (LED), phosphors and quantum dots (QD) have gained considerable scientific interest due to their broad range of applications especially in lighting and display technologies. These phosphors and QDs are used to convert the blue light of the LEDs into white in general lighting. On the other hand, in display systems, they are used to generate red and green bands. There are different application methods such as on-chip and remote configurations. In this study, we concentrate on remote phosphor and QD backlight configurations where the light conversion is done away from the chips. In our display designs, we used GaN based blue LED lateral chips as an excitation source, on the other hand, light conversion layers were placed in backlight units as a thin film for the emission of green and red bands. The mixing ratios of these composite layers were arranged to match the emission spectrum of the blue LEDs and the light conversion layer to the colour filters of the LCD, so that the green, blue, and red bands efficiently widens the colour space. The results were also compared with the on-chip phosphor arrangements.

1. Introduction

Lighting systems based on blue GaN LEDs require a light conversion mechanism in order to produce other colours. This is generally done at the packaging level by covering the LED die with a colour converting layer which is usually a phosphor containing encapsulant. The composition of this encapsulant is critical in determining the spectral and structural quality of the LED chip [1, 2, 3, 4]. In the early LED applications, these on-chip phosphor conversions became quickly popular and are still widely used in general lighting. However, it was later revealed that placing this conversion layer to a remote position away from the LED die improves the device reliability and lifetime [2, 3, 4]. Analogous approaches are also being used in the implementation of quantum dot containing films into the backlight units for widening the colour gamut of the LCD TVs [5, 6, 7].

Colour gamut is one of the quality parameters of display industry which defines the colour space generated by the display unit. Since the LCD based LED displays can be technically described as the combination of the backlight unit (BLU) and the liquid crystal panel (LCP), the optical design

IOP Conf. Series: Journal of Physics: Conf. Series **961** (2017) 012010 doi:10.1088/1742-6596/961/1/012010

characteristics of both these parts should be compatible to get higher performance from the display system [6].

In LED based LCDs, the liquid crystal cells (LC) is illuminated by a backlight unit (BLU). These BLUs are designed as either Direct Light Emitting Diode (D-LED) or Edge Light Emitting Diode (E-LED) configurations according to the position of the LEDs in the BLU. BLUs also include several critical optical components like prism sheets, polarizer sheets, and additional diffuser sheets to improve the homogeneity and brightness (Figure 1).

E-LED designs can be very thin and curved, but their thermal management is difficult due to fact that the light source is located in a narrowly defined area. To solve this issue, the light source is positioned on high cost metallic compounds. Another issue is the distribution of the LED output throughout the BLU. This is done by using a light guide plate (LGP) in which the LED output is propagated by total internal reflection (TIR) (Figure 2).

In D-LED designs, an LGP is not necessary since the LEDs are already positioned in the reflector panel homogeneously and directly. However, this arrangement leads to thicker BLU designs. In addition, the LEDs used in D-LED configurations have a light scattering limit. For this reason, there exists a possibility in case the light output does not smoothly distributed, so the individual LEDs may become visible from the front side of the panel. This problem is generally solved by introducing diffusive elements to the BLU (Figure 3).

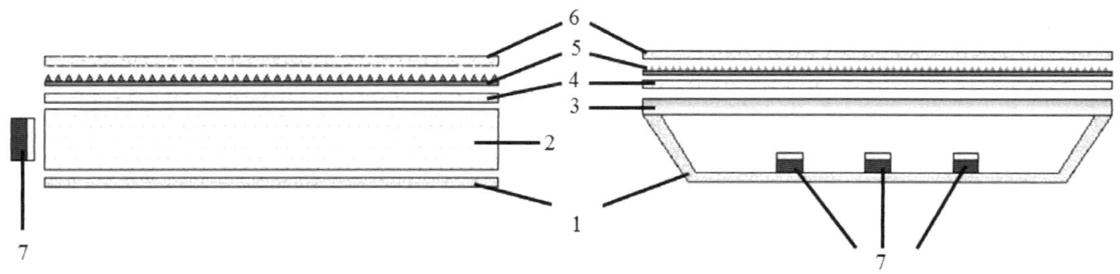

Figure 1. BLU structure of E-LED and D-LED systems. [1]Reflector sheet, [2]Light Guide Plate, [3]Diffuser Plate, [4]Diffuser Sheet, [5]Brightness Enhancement Film, [6]Reflective Polarizer Film, [7]White LED (Blue chip with yellow phosphor)

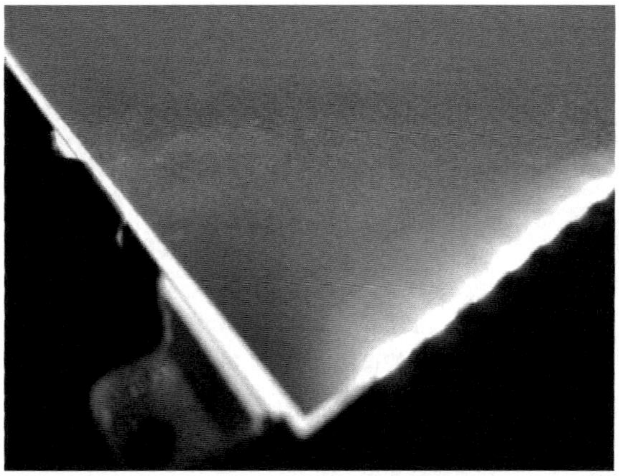

Figure 2. Light guide plate

Figure 3. Diffuser plate

2. Wide Colour Gamut solutions based on phosphor sheets

The display quality is generally defined by central luminance, colour coordinates (CIE-x & CIE-y), gamut, and the brightness uniformity. These parameters depend on both BLU and LCD characteristics.

BLU design is critical in brightness and gamut. The optical components of the BLU are generally selected to maximize the light output of the LEDs for the LC panel, while the spectral output should match the LC panel characteristics. Typical LC panel structure depicted in Figure 4 includes blue, green and red colour filters. The combination of BLU spectrum and CF determines the display output gamut and brightness. For this reason, the colour conversion layer and colour filters inside the LC panel should be optically and spectrally compatible. In phosphor based solutions, this is practically done by determining the phosphor blend according to the CF.
To convert the blue light of the LEDs to CF compatible green and red colour bands effectively, narrow band emitting colour converters provide better solutions, since the combined RGB output of the BLU passes through the CF optimally. However, in most displays, single YAG phosphor is used, which results in narrowing of the gamut and loss of brightness. In addition, on-chip light conversions results in extra heat and light loss due to the phosphor containing polymeric encapsulant on top of the LED die [2, 3]. These issues can be solved by locating the phosphor to a remote position (Figure 5) [1, 2, 3].

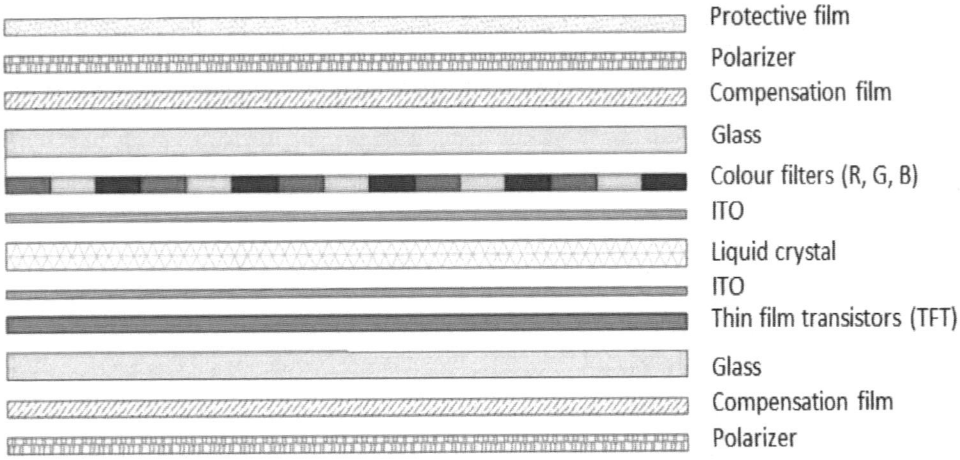

Figure 4. Liquid crystal cell structure

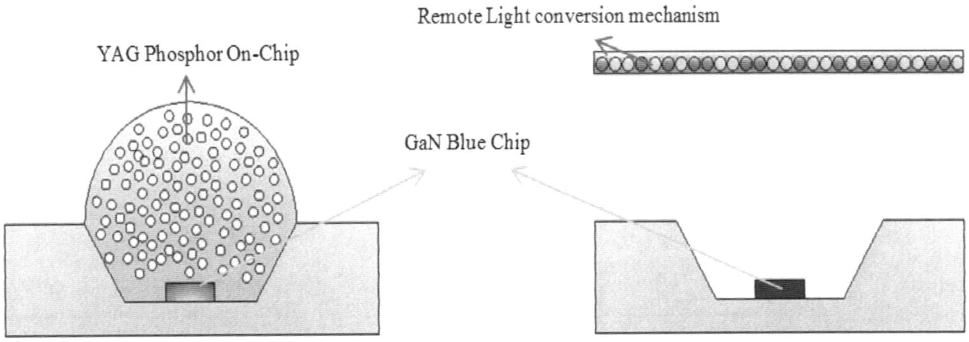

Figure 5. LED on chip vs remote light conversion mechanism

3. Phosphor Blending and BLU Design

We have used green and red emitting phosphors in producing colour conversion layers. In addition, we have compared the performance of these layers with commercially available phosphor and quantum dot containing sheets.

The BLU structure used in the experiments contains lateral blue LED chips which have peak emission wavelengths of 444 nm and FWHMs of 19 nm (Figure 6). The optical layers of this BLU design consists of a diffuser plate (EML-R35A), a light conversion layer (RLCM), a prism on prism sheet (POP 6), and an upper diffuser sheet (KDD-188T2). The optical properties of these layers are given in Table 1. These components play significant role on directing the blue light towards the light conversion layer.

The light conversion layer can be placed as a thin-film coating on the diffuser plate or as a roll-to-roll (R2R) film. The advantage of coating the diffuser plate is to prevent the light loss due to barrier films. However, R2R films are preferred for mass manufacturing.

To prepare phosphor containing encapsulants, we preferred optical silicones (Dow Corning OE-7620) and LED phosphors (Mitsubishi Chemical, $CaAlSiN_3$:Eu and $(Si,Al)_3(O,N)_4$:Eu). The composite mixture was poured down on a diffuser plate and flattened by Dr. Blade applicator. Then, the coated plate was heat cured at 80°C for two hours. The emission properties these phosphor blends under the excitation of a blue LED are given in Table 2.

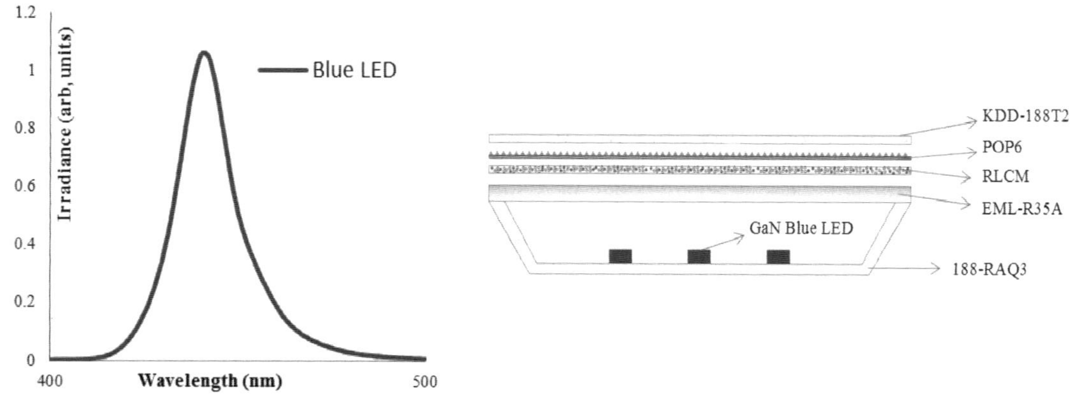

Figure 6. BLU structure of remote light conversion mechanism

Table 1. Optical properties of light conversion mechanism BLU structure

Product Type	*Product Code*	*Thickness (mm)*	*Reflectivity 555nm (%)*	*Haze (%)*	*Transmittance (%)*	*Used Side Gloss (%)*
SMART*	POP6	0.33	N/A	99	3.00	N/A
DP	EML-R35A	1.5	N/A	99	33.00	N/A
RFL	188-RAQ3	0.188	97	N/A	N/A	20.00
UDS	KDD-188T2	0.215	N/A	85	95	N/A

Table 2. The properties of β-SiAlON & CASN phosphors

Chemical Formula	Peak Wavelength (nm)	FWHM (nm)	d50 (μm)
$CaAlSiN_3$:Eu	659	88	16
$(Si,Al)_3(O,N)_4$:Eu	540	52	23.3

4. Experiments and Results

In our display design, we used the BLU structure described above and an LC panel by AUO (65″ UHD T650QVN06) which has a transmittance value of % 4.84. The colour conversion is done by phosphors coatings (0.65mm β-SiAlON & CASN & 0.55mm β-SiAlON & CASN), phosphor R2R sheets, quantum dot R2R sheets, and phosphor converted LEDs (pcLED) for comparison (Figure 7). These are designed to match the CF of the LC panel to improve the brightness and gamut. Our panel CF spectrum is given in Figure 8. Here, all these layers were illuminated by the same optical configuration, except the pcLED reference which is measured without the light conversion layer but by using pcLEDs and BLU with the same optical structure.

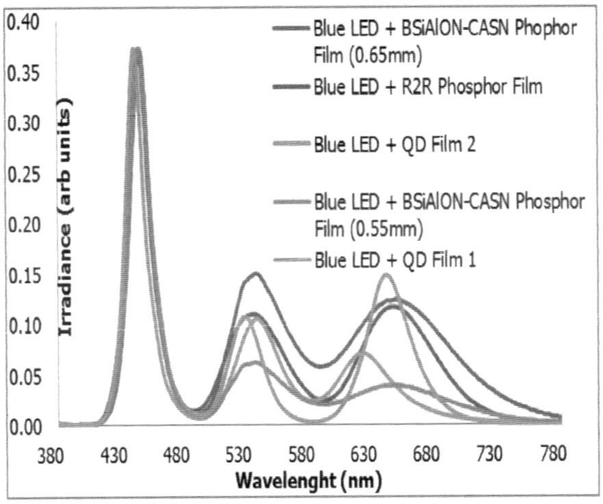

Figure 7. The BLU spectral measurements of β-SiAlON & CASN phosphor blend for different thicknesses (0.55mm & 0.65mm), R2R film, and QD films with respect to white LED

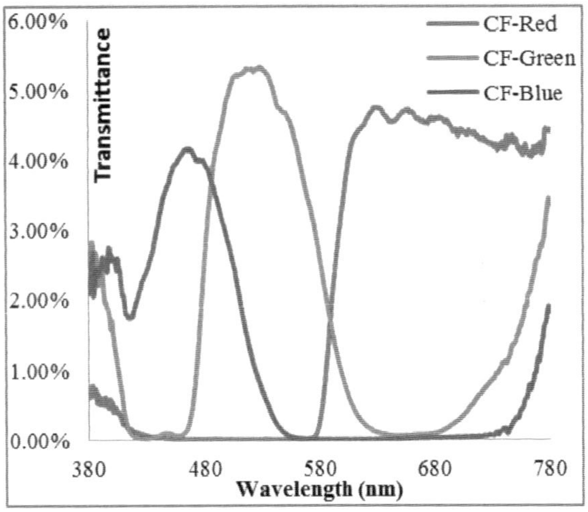

Figure 8. CF spectrum of 65" AUO UHD LCC

The colour coordinates and gamut values were measured by using a spectroradiometer (Topcon, SR-3A) at 50cm away from the center of the display. The measurements were performed after a warm-up period of one hour. The results of DCI-P3 coverage and gamut calculations are presented in Figure 10 and Table 3 below.

Table 3. Colour gamut and coordinates values of different light conversion mechanisms

Light Conversion Mechanism	*R (u'; v')*	*G (u'; v')*	*B (u'; v')*	*DCI-P3 (Area %)*	*DCI-P3 (Coverage %)*
R2R phosphor film	0.515; 0.516	0.098; 0.569	0.178; 0.160	102	96
Beta SiAlON+CASN layer 0.65mm	0.495; 0.521	0.110; 0.572	0.175; 0.172	100	93
Beta SiAlON+CASN layer 0.55mm	0.487; 0.514	0.105; 0.565	0.193; 0.117	102	91
QD Film 1	0.543; 0.511	0.072; 0.571	0.185; 0.149	118	95
QD Film 2	0.486; 0.519	0.104; 0.568	0.187; 0.134	99	93
On-Chip yellow phosphor	0.450; 0.524	0.119; 0.568	0.182; 0.153	82	82
DCI-P3 Colour Gamut Standard	0.496; 0.526	0.099; 0.578	0.175; 0.158	100	100

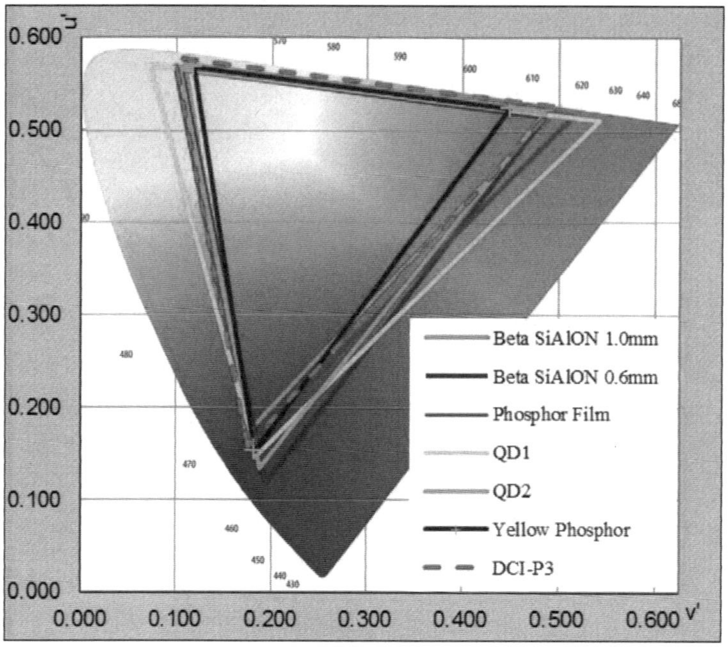

Figure 9. Colour gamut area of β-SiAlON & CASN phosphor blends for different thicknesses (0.55mm & 0.65mm), R2R film, and QD films with respect to pcLEDs at CIE 1976 colour space

5. Conclusion

We have performed several light conversion layer types to improve the colour gamut of LCD based displays. Here we have shown that the remote light conversion mechanism is not only effective in expanding the colour space but also useful in thermal management. In addition, it has a direct effect on the lifetime of the system. When it was arranged to match the CF spectra of the LC panel, it also increase the brightness of the display unit. On the hand, these layers can also be prepared by using QDs. However, for heavy metals free QDs, the gamut and intensity values lower than both the phosphor blends and Cd-based QD films. For this reason, phosphor based light conversion layer is expected to be in use for wide colour gamut solutions.

6. References

[1] E. F. Schubert, "*Light-emitting diodes,*" Cambridge University Press, 2006

[2] N. Y. Narendran, et al. "*Extracting phosphor-scattered photons to improve white LED efficiency,*" Phys. Stat. Sol. (a) 202 (6): R60-R62 2005

[3] T. Güner, D. Köseoğlu, Mustafa M. Demir, "*Multilayer design of hybrid phosphor film for application in LEDs,*" Optical Materials 2016 vol **60** pages 422–430

[4] D. Lu & C.P. Wong (Eds.), "*Materials for Advanced Packaging,*" Chapter 18 pp 629 – 680 Springer 2009

[5] J. Van Derlofske & A. Lathrop, "*Quantum dot film lets LCDs express 50% more color,*" Laser Focus World, Vol. 50, Iss. 4, pp. 39-42, April 2014

[6] H. Chen, J. He, and S. Wu, "*Recent Advances on Quantum-Dot-Enhanced Liquid-Crystal Displays*", IEEE J. of Selected Topics in Quantum Electronics, Vol. 23, Iss. 5, Sep/Oct 2017

[7] S. Abe, et al. "*Hybrid remote quantum dot/powder phosphor designs for display backlights,*" Light: Science & Applications (2017) 6, e16271, 2017

Acknowledgements

This work is supported by the Scientific and Technological Council of Turkey (TUBITAK).

IOP Conf. Series: Journal of Physics: Conf. Series **961** (2017) 012011 doi:10.1088/1742-6596/961/1/012011

Thermal Dynamics of Xanthene Dye in Polymer Matrix Excited by Double Pulse Laser Radiation

Ilia Samusev[1], Rodion Borkunov[1], Maksim Tsarkov[1], Elizaveta Konstantinova[1,2], Yury Antipov[2], Maksim Demin[1], Valery Bryukhanov[1]

[1]Research and Education Center "Fundamental and Applied Photonics. Nanophotonics", Center for Functionalized Magnetic Materials, Immanuel Kant Baltic Federal University, Nevsky Str. 14, Kaliningrad, Russia;
[2]Kaliningrad State Technical University, Sovetsky Ave. 1, Kaliningrad, Russia

is.cranz@gmail.com

Abstract. Double-pulse laser excitation of the eosin and silver nanoparticles embedded into polymer media is known to be a method of electronic-vibrational energy deactivation kinetic process information obtaining and polymer thermal dynamics investigation. We have studied the vibrational relaxation processes in dye molecules (eosin) and nanoparticles in polyvinyl alcohol after two time-shifted laser pulses with fast and delayed fluorescence kinetics study. In order to simulate thermal and photophysical processes caused by double photon excitation, we solved heat transfer and energy deactivation differential equations numerically. The simulation allowed us to obtain the value of heat conductivity coefficient of polymer matrix.

1. Introduction

Fluorophors and silver nanoparticles (NP) embedded into polymer media double-pulse laser excitation (DPLE) is known to be a method for molecule energy deactivation information obtaining and polymer thermal dynamics investigation [1–8]. Nowadays, DPLE in VIS/IR wavelength range can be considered as a tool for detailed investigation of fluorescence resonance energy transfer (FRET) in donor-acceptor molecular systems [9], electron-phonon interaction study [10–12] and fabrication of ablative NP [13], which can take part in plasmon energy transfer [14]. We aimed to consider vibrational relaxation processes in dye molecules (eosin) and NP embedded into polymer matrix (polyvinyl alcohol, PVA) after two time-shifted laser pulses with fast and delayed fluorescence (DF) kinetics.

2. Experiment

The experimental setup is illustrated in Figure 1. Eosin molecules in polyvinyl alcohol (PVA) DF and phosphorescence (Ph) after double laser pulse radiation (Vis and IR) were measured by photon counting unit (Hamamatsu C8855-01). A metallic plate with 5 mm diameter hole was covered by the PVA film. All measurements were carried out at a room temperature and a pressure of 10^{-4} torr.

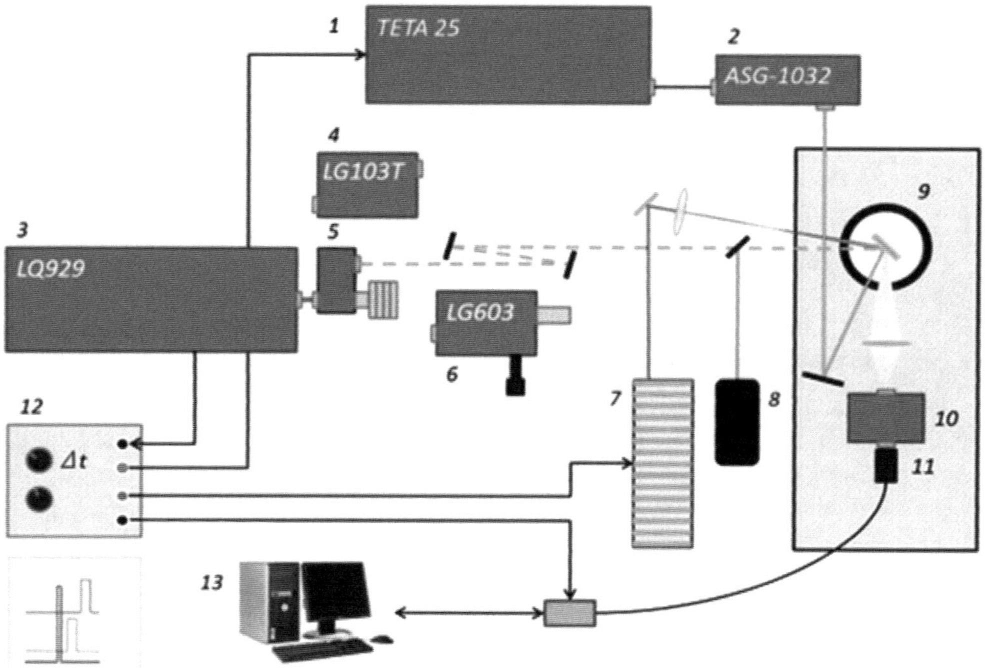

Figure 1. Experimental setup: 1-2: fs-laser unit Teta-25/30 (τ = 60 fs, W = 0.12 μJ, λ = 515 nm); 3-6: YAG Nd^{3+} laser unit LQ929 (Solar Laser Systems); 7: CO_2 laser unit (steady/pulse regimes, τ = 50÷200 ms, λ = 10 μm); 8: Semiconductor laser unit ASG-1032 (W = 50 mW, λ = 532 nm); 9: cryostat Optistat VN-V; 10: monochromator ML44; 11: PMT/Photon counting unit C8855-01 (Hamamatsu C8855-01); 12: 2-channel pulse generator G-56; 13: PC.

3. Results and Discussion

Figure 2 illustrates the eosin (0.5 mM) fluorescence spectra at stationary (green) and 50 ms delayed IR pulse (red) excitation. Eosin DF and Ph in PVA film were studied at 555 and 680 wavelengths respectively after LQ929 YAG:Nd^{3+} laser excitation (λ = 532 nm, W = 15÷20 J, τ = 11 ns) and IR CO_2 laser pulse (λ = 10,6 μm, $P \approx 20$ W). The IR-pulse was time-shifted and irradiated the film within eosin DF decay time range.

It is known that fluorophors' DF intensity depends on the interconversion transitions efficiency ($k_{ic} \sim 10^8$ s) [15] and concentration of molecules in triplet state. The polymer matrix heating by the IR-pulse causes eosin triplet state electron-vibration levels population increase ($\Delta\nu_{ST}$ = 3133 cm^{-1}) [16].

Eosin DF intensity decay after IR pulse is shown in Figure 3. The DF kinetics lifetimes before and after the IR laser pulse 0.017 and 0.050 s respectively. It can be seen from the Figure 3 that there is a 20% local increase of DF intensity and lifetime directly after the CO_2 laser pulse.

Assuming that such enhance is caused by eosin singlet molecules population increase due to interconversion process, it is possible to estimate the temperature leap right after the IR laser pulse:

$$\Delta T = T_0\left[\frac{1}{(1-k_B T/\Delta E_{ST})\ln I_1^{DF}/I_2^{DF}} - 1\right], \quad (1)$$

where T_0 denotes an ambient temperature; k_B denotes the Bolzman constant; ΔE_{ST} denotes the singlet-triplet energy splitting; I_1^{DF} and I_2^{DF} are the eosin DF intensities right before and after the IR pulse. The temperature leap was estimated to be 4 K.

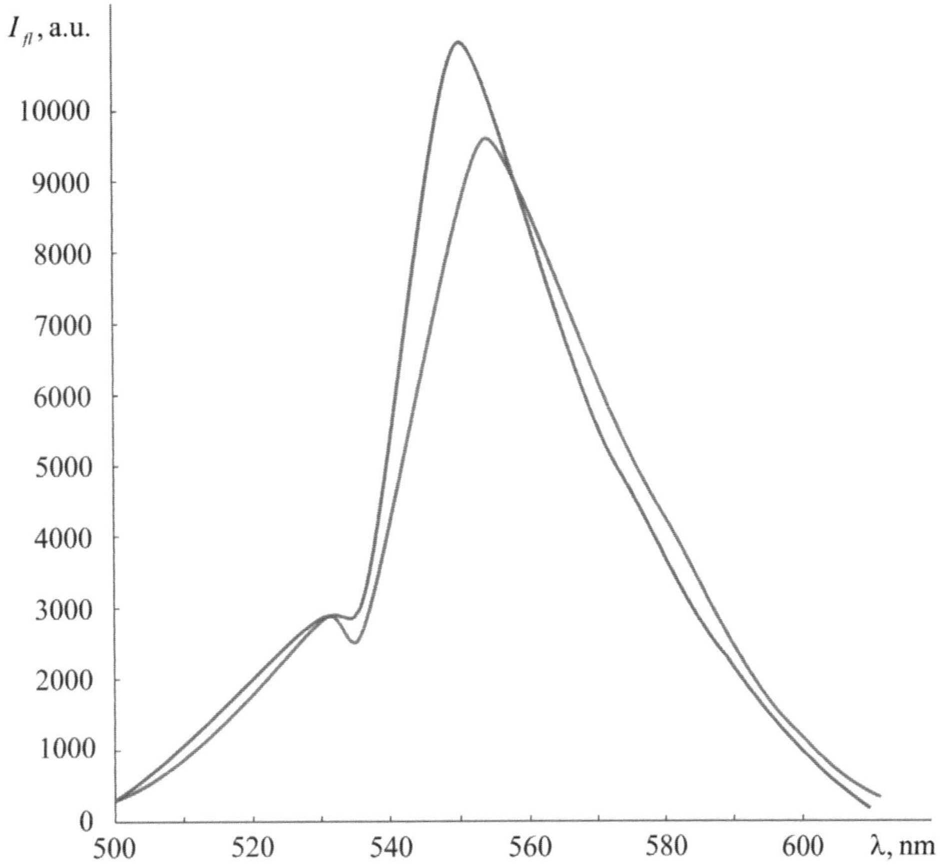

Figure 2. Eosin (0.5 mM) fluorescence spectra at stationary (green) and 50 ms delayed IR pulse (red) excitation.

In the second series of the experiment, PVA films with eosin were excited by stationary semiconductor laser ASG-1032 radiation (λ = 532 nm, W = 50 mJ) and CO_2 laser time-shifted (50 ÷ 200 ms) IR pulse (λ = 10,6 μm, $P \approx$ 20 W), which allowed us to study the kinetic processes eosin singlet and triplet states population decay after IR heating.

DF kinetics was exponential with a lifetime of 4.7 s. Simulation of the eosin DF intensity decay in PVA allowed us to obtain the quantitative characteristics of the thermal energy transfer in the polymer matrix.

Heat equations with the corresponding initial and boundary conditions for the studied system were solved [17]:

$$\begin{aligned}
&\frac{\partial T}{\partial t} = a^2 \frac{\partial^2 T}{\partial x^2};\ t > 0,\ -R < x < R; \\
&T(x,0) = \Theta(x); \\
&-\lambda \left.\frac{\partial T}{\partial x}\right|_{x=-R} + \alpha[T_0 - T(+R,t)] = 0; ' \\
&-\lambda \left.\frac{\partial T}{\partial x}\right|_{x=+R} + \alpha[T_0 - T(-R,t)] = 0,
\end{aligned} \tag{2}$$

where R denotes the PVA film halfwidth; a and λ are the temperature conductivity and heat transfer coefficients, respectively; α denotes the heat emission constant.

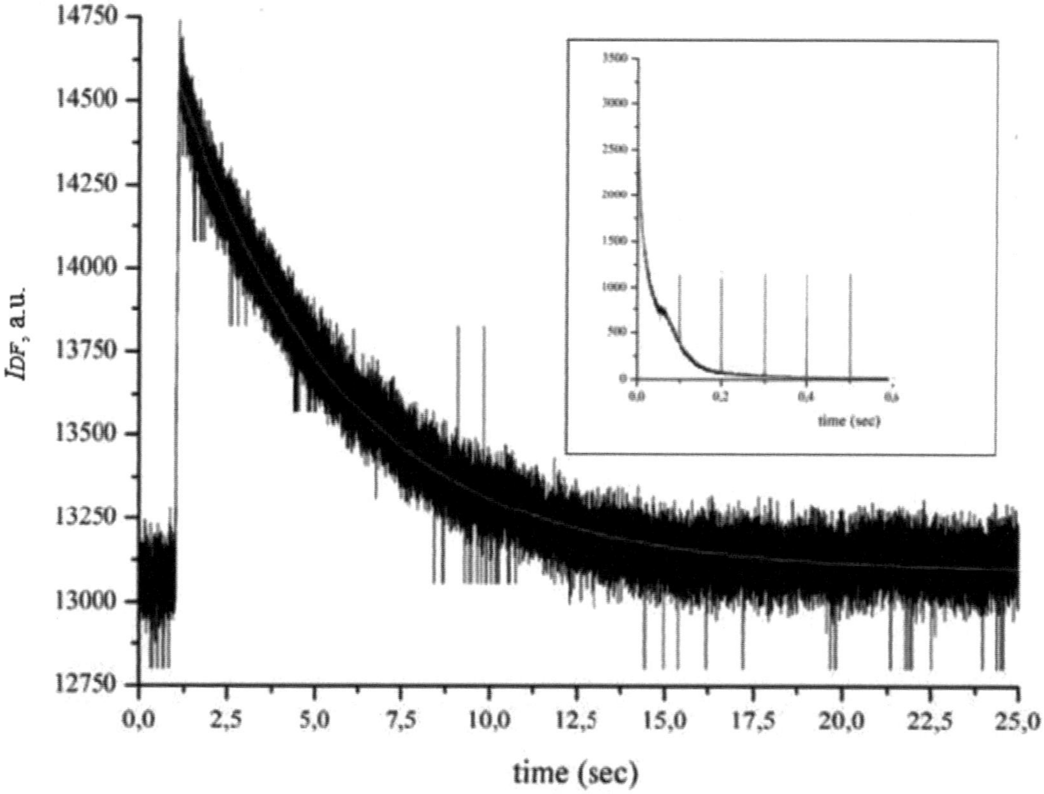

Figure 3. Eosin DF intensity decay after IR pulse.

Eq. (2) was solved for the temperature leap ΔT by variables separation method: $\Delta T \sim \psi(x)\exp(-a\mu^2 t/R^2)$ where μ = 0.0316 [5]. The value of a was found to be 0.243 · 10^{-6} m^2/s which coincides with literature data for heat characteristics of PVA.

PVA film doped with silver NP heat transfer characteristics were investigated in the third series of our experiment. Silver NP ablation was performed by means of fs-laser unit Teta 25/30 (Avesta) (τ = 60 fs, W = 0.12 μJ, λ = 515 nm) in an aqueous medium. The silver target was placed in a cuvette with a liquid on the rotary platform. The laser beam was guided and focused on the surface of the target. Silver NP sizes were measures by photon correlation spectroscopy unit Photocor-C with 432 nm-laser.

NP average radius was 32 nm. Silver NP clusters with low silver concentrations can serve as heat point centers with a temperature much higher that the polymer matrix. The temperature field dynamics can be simulated by heat transfer equations with spherical Greene function. We solved the following heat transfer problem for above-mentioned system:

$$\begin{aligned}
&\frac{\partial T}{\partial t}=\frac{a}{r^2}\left\{\frac{\partial}{\partial r}\left[r^2\frac{\partial T}{\partial r}\right]+\frac{\partial}{\partial \vartheta}(1-\vartheta^2)\frac{\partial T}{\partial \vartheta}+\frac{1}{1-\vartheta^2}\frac{\partial^2 T}{\partial \varphi^2}\right\};\\
&T(r,\vartheta,\varphi,0)=f(r,\vartheta,\varphi);\\
&\lambda\frac{\partial T}{\partial r}=\alpha_R[T(R,\vartheta,\varphi,t)-T_0];\\
&\vartheta=\cos\theta;\ 0\le\varphi\le 2\pi;\ -\frac{\pi}{2}\le\theta\le\frac{\pi}{2},
\end{aligned} \quad , \qquad (3)$$

where T denotes the cluster's temperature, R denotes the radius of the cluster.

The standard solution of the eq. (3) can be expressed as

$$T(r,t)=\frac{Q}{8\pi rRc\rho(\pi at)^{1/2}}\left\{\exp\left[-\frac{(r-R)^2}{4at}\right]-\exp\left[-\frac{(r+R)^2}{4at}\right]\right\}, \qquad (4)$$

where Q denotes the heat power after IR laser pulse; c and ρ are the heat capacity and density of the cluster. The estimated value of a for the system with silver NP as heat centers was $0.096 \cdot 10^{-6}$ m^2/s.

4. Conclusion

Thus, using double-pulse laser excitation (Vis and IR) of the fluorophor electron-vibrational states in polymer films, it is possible to obtain the information about thermal processes in homogeneous polymer films.

Aknowledgements

This work is supported by the Russian Academic Excellence Project ("Project 5-100") of the Ministry of Education and Science of the Russian Federation.

References

[1] Letokhov V S. 1976 *Uspehki Fizicheskikh Nauk* **118** 2 149–299
[2] Brumbargh D V., Kenny J E., Levy D H 1983 *Journal of Chemical Physics* **78** 3415–34
[3] Miyazaki Y, Inokuchi Y, Ebata T, Petkovici M 2013 *Chemical Physics* **419** 205–11
[4] Yang F, Jiang L, Wang S, Cao Z, Liu L, Wang M, Lu Y 2017 *Optics and Laser Technology* **93** 194–200
[5] Ciofini D, Striova J, Camaiti M, Siano S 2016 *Polymer Degradation and Stability* **123** 47–61
[6] Takeshita T, Kurata H, Hara M 2017 *Journal of Photochemistry and Photobiology A: Chemistry* **334** 28–35
[7] Komarov V V, Popova A M , Stureyko I O , Shmidt L , Yungklas H 2013 *Physics and Astronomy Series* **5** 3–7
[8] Makarov G N 2017 *Uspekhi Fizicheskikh Nauk* **187** 3 241–76
[9] Odeneye M A 2005 *Physical Chemistry Chemical Physics* **5** 998–1004
[10] Sobolev S L. 2016 *International Journal of Heat and Mass Transfer* **94** 138–44
[11] Du G, Wu Y, Uddin N, Yang Q, Chen F, Lu Y Bian H, Hou X 2016 *Optics Communications*

IOP Conf. Series: Journal of Physics: Conf. Series **961** (2017) 012011 doi:10.1088/1742-6596/961/1/012011

375 54–7
[12] Kayano M, Ebata T, Yamada Y, Mikami N 2004 *Journal of Chemical Physics* **120** 16. 7410–17
[13] Khalil A A 2015 *Optics and Laser Technology* **75** 105–14
[14] Klimov V V 2009 *Nanoplasmonics* (Moscow: Fizmatlit) p 480
[15] Gorbatsevich S A *Spectroscopy of Intermolecular Interactions. Nonlinear effects* 2004 (Minsk: Elektronnaya Kniga BSU)
[16] Letuta S N (2003) Photoprocesses with polyatomic molecules' higher electronic-excited states *Doc. Sci. Dissertation Autoref.* (Moscow: MSU)
[17] Ponomarev S V, Mishchenko S V, Divin A G (2006) *Theoretical and Practical Aspects of Thermophysical Measurements* (Tambov: TSU)

IOP Conf. Series: Journal of Physics: Conf. Series **961** (2017) 012012 doi:10.1088/1742-6596/961/1/012012

Dirac-Weyl points' manipulation using linear polarised laser field in Floquet crystals for various Graphene superlattices

S.A. Alfadhli[1,2], S.E. Savel'ev[1] and F.V. Kusmartsev[1]
[1]Department of Physics, Loughborough University, Loughborough LE11 3TU, United Kingdom
[2]Department of Physics, University of Tabuk, Tabuk, 71491, Saudi Arabia

Abstract. We investigate the changes in the energy spectrum of the graphene monolayer subjected to linear polarised laser beam and external periodically modulated static field (electric and magnetic). Floquet theory and the resonance approximation are used to analyse the energy spectrum and, in particular, the creation and the destruction of the Dirac-Weyl points. We found that at certain conditions the graphene is transformed into the two-dimensional Weyl metals, where each of the two original graphene Dirac cones is split into pairs of the Weyl cones. We also show that altering the laser's beam incidence(tilting) angle may lead to appearing and disappearing of the pairs of Weyl points, the opening gap in the spectrum, and its efficient manipulation.

1. Introduction

The two-dimensional nature of graphene discovered in 2004 has enlightened the importance of potential applications of the 2D crystals [1] [2] [3] [4]. The gapless energy structure of graphene allows the electrons to travel freely with a very high mobility [5] and small electrical resistance [6]. However, the optoelectronic application of graphene requires an opening energy gap in the Dirac spectrum. [7] [8] [9] [10] [11]. Specifically, for a transistor construction and its various application the gap should be opened and closed with very high gigahertz (GHz) or even terahertz (THz) frequency and that should be controlled by varying external circuit parameters [12] [13] [14]. One of the promising gap controlling techniques is a generation of periodic potential, i.e. the creation of the so-called graphene superlattices. Such devices may have a very high electric mobility [15] [16], that is a prerequisite for the Bloch oscillations [17] [18] [19] [20]. From other side that can provide a strong potential for numerous GHz and THz practical applications [21] [22] [21]. An analogy was found between this costly-delicate method and between graphene placed in laser fields [23] [24] [25] [26] [27]. The laser field provide very nontrivial changes with materials and especially with graphene and may be used to test novel topological phases. For example, an application of the laser field to epitaxial graphene grown on SiC subjected even to very weak magnetic field transforms it into topological insulator with very nontrivial topological phases [28]. Moreover, an application of both constant electrical and inhomogeneous or periodic magnetic field can create in graphene snake states which is a feature of another class of topological insulator, the topological classification, Class AII, $Z_2 = 1$ [29]. The similar situation arises when we apply a constant magnetic and periodic electrical field. In this case single layer graphenes have one-dimensional domain walls that can arise in the places where the electrical or magnetic field changes sign. There arise snake states induced by orbital quantization in the spatially varying perpendicular magnetic field. Thus the periodic potential, either electrical or magnetic form a superlattice of domain walls. Each domain wall may be viewed as a p-n junction, which has topologically protected states, where there are topologically protected orbital currents.

These confined electronic states may provide configurable transport channels for charge and spin. Transport through these domain wall channels and switching from the edge to the internal pathways can be in principle controlled by an electrostatic gate and the laser field. The periodic electrical potential can be provided e.g. by grain boundaries. That interfacial states on zero-angle grain boundaries are actually topologically protected zero modes that are protected by a novel nonlocal chiral symmetry of the graphene Hamiltonian near its charge neutrality point or the Dirac point. The significance of this nonlocality was shown for the first time in Ref. [29]. It was shown that the spectra for laterally confined states at p-n junctions and snake states for modulated B fields with uniform chemical potential are gauge equivalent representations of the same problem. When expressed in a Nambu particle-hole doubled representation the two problems are formally interconverted by a position-dependent particle-hole transformation. There ballistic transport through a gate defined p-n junction can be configured as a gate-tunable valley filter. Thus the system is very reach has numerous practical applications has a nonlocal symmetry and therewith is analogous to Weyl semi-metals [30], where Dirac-Weyl points may appear and disappear in pairs when the spatial inversion is broken. All these analogies are described in the recent review [31]. Having that in mind in this paper, we theoretically investigate the energy spectrum in single layered graphene superlattices controlled by the laser field, when simultaneously an external either electric or magnetic field is applied. Since the system has a very rich symmetry the application of the laser field can influence the number and the position of the Dirac point and in general change the energy spectrum. Indeed, we show that the spectra and the current flow of Dirac electrons in graphene can be controlled by altering the laser beam intensity and the incidence angle with respect to the graphene plane.

2. Modelling approach:

To describe the motion of electrons in graphene subjected to periodic time dependent and electro-magnetic field we use the 2D Dirac equation in the form [20]:

$$\{\vec{\sigma}.[\vec{p}\text{-}\vec{A}(t)\text{-}\vec{A}^H(x)] + U(x)\}\,\psi_{A,B} = \frac{i\partial\psi_{A,B}}{\partial t} \tag{1}$$

where $\vec{A}(t)$ is the part of the vector potential associated with the time dependent electric field, $\vec{A}^H(x) = \left(\vec{A}^H(x), \vec{A}^H_y(x)\right)$ is its magnetic part, $U(x)$ is a static electric field, and $\psi_{A,B}$ is the electron wave function represented by two components vector $\psi = \begin{pmatrix}\psi_A\\ \psi_B\end{pmatrix}$ referring to the electrons in the two triangular sublattices A and B, respectively. The expression (1) is written in units of $e = \hbar = v_F = 1$. Floquet theory can be used to analyse an evolution of electron wave function in a periodic time dependent potential. It gives a solution in the form a "Bloch function" in the time domain: $\Psi(x,t) = e^{-i\varepsilon t}\Phi(x,t)$, where the function $\Phi(x,t)$ is periodic in time, $\Phi(x,t) = \Phi(x,t+T)$, and the value ε is the quasienergy or the Floquet energy. In the resonance approximation [20], where we are interesting only in the low-energy spectrum in a vicinity of the Dirac point of the graphene spectrum the wave function can be written as (e.g. see the Ref.[24], for a detail):

$$\psi_{A,B} = e^{-i\varepsilon t+ik_x x+ik_y y}\left[\psi^{++}_{A,B}\,e^{\frac{i\omega t}{2}+\frac{i\mu x}{2}} + \psi^{+-}_{A,B}\,e^{\frac{i\omega t}{2}-\frac{i\mu x}{2}} + \psi^{-+}_{A,B}\,e^{\frac{-i\omega t}{2}+\frac{i\mu x}{2}} + \psi^{--}_{A,B}\,e^{\frac{-i\omega t}{2}-\frac{i\mu x}{2}}\right] \tag{2}$$

where ω and μ are temporal frequency and spatial modulation inverse period of laser field and spatially-periodic static field (either electric or magnetic). We apply a linearly polarized laser field :

$$\vec{A}(t)_x = A_0 cos(\theta)cos(\omega t)\,;\, \vec{A}(t)_y = A_0 sin(\theta)cos(\omega t) \tag{3}$$

with the amplitude A_0 and polarization angle θ, which is the angle between the linear polarised laser field and the static field. First, we consider a case when magnetic part of the vector potential has the form, where $\vec{A}^H(x)_x = 0, U(x) = 0$. Then, we apply a static electric field oriented in the graphene plane. The field is associated with an electric potential of the form: $U(x) = U_0 cos(\mu x)$, where U_0 is the amplitude and μ is the value, inverse to the spatial modulation period, $d = 2\pi/\mu$. In the considered case the electric field will be always in plane with the graphene sheet. Substituting the wave function and these potentials into the 2D Dirac equation (1) we obtain the following expressions:

$$i\frac{d}{dt}\begin{pmatrix}\phi_A\\ \phi_B\end{pmatrix} = \begin{pmatrix} U(x) & [\mathrm{p_x}\text{-}\mathrm{A_x}(t,x)]\text{-}i[\mathrm{p_y}\text{-}\mathrm{A_y}(t,x)] \\ [\mathrm{p_x}\text{-}\mathrm{A_x}(t,x)] + i[\mathrm{p_y}\text{-}\mathrm{A_y}(t,x)] & U(x)\end{pmatrix}\begin{pmatrix}\phi_A\\ \phi_B\end{pmatrix} \quad (4)$$

where $\mathrm{A(t,x)} = (\mathrm{A_x}(t,x), \mathrm{A_y}(t,x)) = ((\mathrm{A_x}(t)\text{-}\mathrm{A_x^H}(x)),\ (\mathrm{A_y}(t)\text{-}\mathrm{A_y^H}(x)))$. By substitution of the expression for the truncated form of the wave function and neglecting the terms associated with higher order harmonics, we obtain the set of eight algebraic homogeneous equations, arranged in the form

$$\begin{bmatrix}
\frac{\omega}{2} & k_x - ik_y + \frac{\mu}{2} & C_nA_0 & 0 & C_mA_0^H & -\left(\frac{A_0}{2}\right)e^{-i\theta} & 0 & 0 \\
k_x + ik_y + \frac{\mu}{2} & \frac{\omega}{2} & C_mA_0^H & C_nA_0 & -\left(\frac{A_0}{2}\right)e^{i\theta} & 0 & 0 & 0 \\
C_nA_0 & C_mA_0^H & \frac{\omega}{2} & k_x - ik_y + \frac{\mu}{2} & 0 & 0 & 0 & -\left(\frac{A_0}{2}\right)e^{-i\theta} \\
C_mA_0^H & C_nA_0 & k_x + ik_y + \frac{\mu}{2} & -\frac{\omega}{2} & 0 & 0 & -\left(\frac{A_0}{2}\right)e^{i\theta} & 0 \\
0 & -\left(\frac{A_0}{2}\right)e^{-i\theta} & 0 & 0 & -\frac{\omega}{2} & k_x - ik_y + \frac{\mu}{2} & C_nA_0 & C_mA_0^H \\
-\left(\frac{A_0}{2}\right)e^{i\theta} & 0 & 0 & 0 & k_x + ik_y + \frac{\mu}{2} & -\frac{\omega}{2} & C_mA_0^H & C_nA_0 \\
0 & 0 & 0 & -\left(\frac{A_0}{2}\right)e^{-i\theta} & C_nA_0 & C_mA_0^H & -\frac{\omega}{2} & k_x - ik_y + \frac{\mu}{2} \\
0 & 0 & -\left(\frac{A_0}{2}\right)e^{i\theta} & 0 & C_mA_0^H & C_nA_0 & k_x + ik_y + \frac{\mu}{2} & -\frac{\omega}{2}
\end{bmatrix}
\begin{bmatrix}\psi_A^{++}\\ \psi_B^{++}\\ \psi_A^{+-}\\ \psi_B^{+-}\\ \psi_A^{-+}\\ \psi_B^{-+}\\ \psi_A^{--}\\ \psi_B^{--}\end{bmatrix}
= \varepsilon
\begin{bmatrix}\psi_A^{++}\\ \psi_B^{++}\\ \psi_A^{+-}\\ \psi_B^{+-}\\ \psi_A^{-+}\\ \psi_B^{-+}\\ \psi_A^{--}\\ \psi_B^{--}\end{bmatrix} \quad (5)$$

Below we present the solution of this equation by setting its determinant equal to zero.

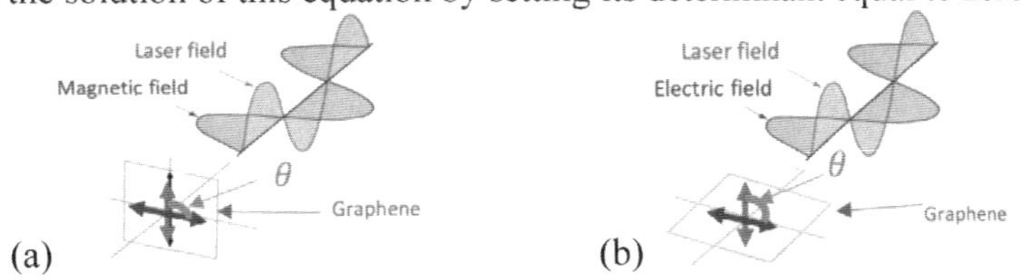

Figure 1. Schematic representation of the orientation set up for the discussed systems where a linearly polarized laser field applied in addition to an external spatially periodically modulated electric or magnetic field. (a) a static, periodically modulated magnetic field is applied perpendicular to the graphene sheet, while the polarisation of the laser field is applied at a chosen incidence angle θ with respect to the orientation of the static field. (b) a static, periodically modulated electric field is located in the graphene plane while a linear polarised laser field is tilted by angle θ with respect to the orientation of the electrical field.

3. Graphene in static magnetic and linear polarised laser field:

In this section we consider the case when the static magnetic field is applied perpendicular to the graphene sheet. Usually we should expect Landau Quantisation here and Landau spectrum consisting of Landau Levels if the spatial modulation period is very large. However, in additional linear polarised laser field and in the static field with the periodic spatial modulation the low energy spectrum from a first glance we see three Dirac points oriented parallel to k_x axes, when the wave vector $k_y = 0$ (see, the Figure 2). It is clear that this "tripling" is related to the spatial modulation of the magnetic field in x- direction which is creating domain walls parallel to the y- axes. However, a more precise inspection of this part of energy spectrum in the vicinity of the zero energy reveals that we have here two Dirac-Weyl points and one highly anisotropic gapless Dirac-like point at $\varepsilon = 0$, corresponding both, $k_x = 0$ and $k_y = 0$. The two Dirac-Weyl points shifted in the k_x direction correspond to the creation of the vortex and the anti-vortex in the energy-momentum space having the form of Dirac cones in the vicinity of the zero energy. When these cones are separated in the momentum space they are topologically protected and we cannot create a gap in the spectrum. But they can annihilate each other when you will be located at the same place and there an energy bandgap can be created. Note that the original Dirac point (in our notation here it is located at K or K' points of the Brillouin Zone but at the values, $k_x = 0$ and $k_y = 0$) is now strongly deformed and probably lost its linear character along k_x direction. The formation of the two Dirac-Weyl singularities in the vicinity of the K and K' points is a result of the dynamical topological phase transition created by an application of the both periodically modulated spatial and dynamical laser field. The position of these two Dirac-Weyl points can be

IOP Conf. Series: Journal of Physics: Conf. Series **961** (2017) 012012 doi:10.1088/1742-6596/961/1/012012

rotated by the changes of the angle between the laser field polarisation and the direction of the applied field of the periodic spatial modulation, see the Figure 3, where for an illustration the calculations have been done with the values of the parameters: $\omega = 1$, $\mu_H = 1$ and $A_0^H = 1$. From this Figure 3 we see that when the polarisation direction changes by $\pi/2$ the position of these two Weyl-Dirac points is perpendicular to their original orientation, when the direction of the laser field polarisation and of the modulated field coincide (see, the Figure 3).

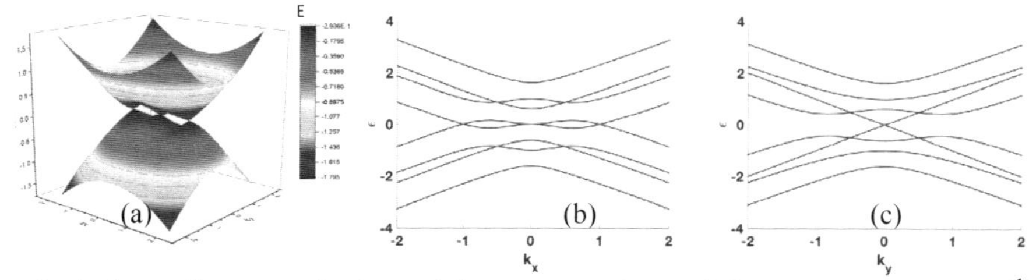

Figure 2. A three-dimensional plot and its cross sections of the energy spectrum $\varepsilon(k_x, k_y)$ of a single layered graphene obtained in the applied linearly polarised laser field together with a spatially-modulated periodic external magnetic field. The spectrum obtained with approximation strictly speaking valid only for the low energy $\varepsilon(k_x, k_y) < 1$. The calculation has been done with the parameters: $\omega = 1$, $\mu_H = 1$ and $A_0^H = 1$ for chosen fields orientations. (a) the spectrum shows two Dirac-Weyl points associated with Dirac cones and one highly anisotropic gapless point at $\varepsilon = 0$, corresponding both, $k_x = 0$ and $k_y = 0$. These two Dirac-Weyl points shifted in the k_x direction correspond to the creation of the vortex and the anti-vortex in the energy spectrum and are topologically protected. Interesting that the original Dirac point, which is in the present notation located at $k_x = 0$ and $k_y = 0$ is now strongly deformed and probably lost its linear character along k_x direction. Note that for graphene this Dirac point correspond to the K and K' points of the Brulluoen Zone. The formation of these two extra Dirac-Weyl cones in the vicinity of the K and K' points is a result of the dynamical topological phase transition created by an application of the both periodically modulated spatial and dynamical laser field. These two Dirac-Weyl points and one Dirac point are aligned along the k_x axes, i.e. at $k_y = 0$. On the Plot (b) we see this cross section while on the Plot (c) we present the cross section of the spectrum taken at $k_x = 0$.

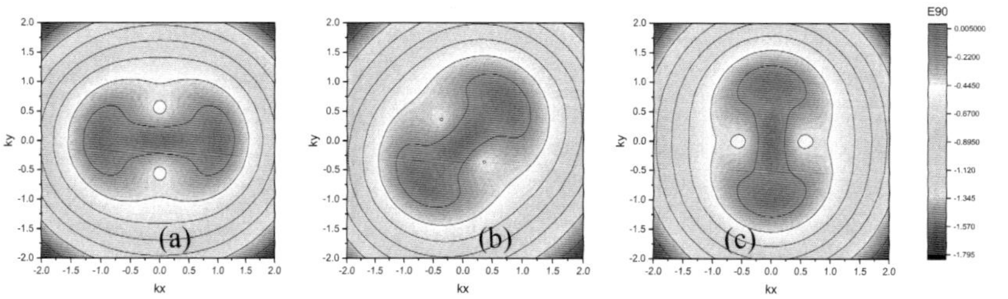

Figure 3. The contour Plot of the low-energy spectrum $\varepsilon(k_x, k_y)$ taken in the vicinity of each K and K' points of the Brillouin Zone for a single layered graphene in the perpendicular periodically modulated magnetic field and a linear polarised laser field tilted at different angles θ with respect to each other. (a) Shows the contour plot of the energy spectrum when $\theta = 0$, the symmetry axes of the spectrum is horizontally oriented parallel to k_x axis which coincides with the polarisation of the electromagnetic laser field. (b) As we rotate the laser field polarisation on $\theta = \pi/4$ or $\pi/2$ angle the Weyl Dirac points of the spectrum also rotates (see, the panels b and c).

4. Graphene in static electric and linear polarized laser field:

If we apply a static periodically modulated electric field and the linearly polarized laser field in the plane of the graphene sheet there are arise a two pairs of the Weyl-Dirac points. In this case when

IOP Conf. Series: Journal of Physics: Conf. Series **961** (2017) 012012 doi:10.1088/1742-6596/961/1/012012

$\theta = 0$ the energy spectrum shows that Dirac cones of graphene split into several mini bands (see, Figure 4-a). The case is very similar to the previous one, the spectrum has two Dirac-Weyl point and one central original Dirac point. However, the elongation of this central point is different, it occurs along k_y axes. All these points are located along the k_x axis and the spectrum seems to be gapless. The 3D plot (see, the Figure 4-b) shows the shape of these Dirac and Weyl cones and that the middle Dirac cone is elliptically shaped. When the angle θ changed from 0 to $\frac{\pi}{2}$, the number and positions of Dirac-Weyl points have changed to four (see, Figure 4-c), indicating a dynamical symmetry breaking phenomenon arising in the band structure with changes of the different laser beam orientations.

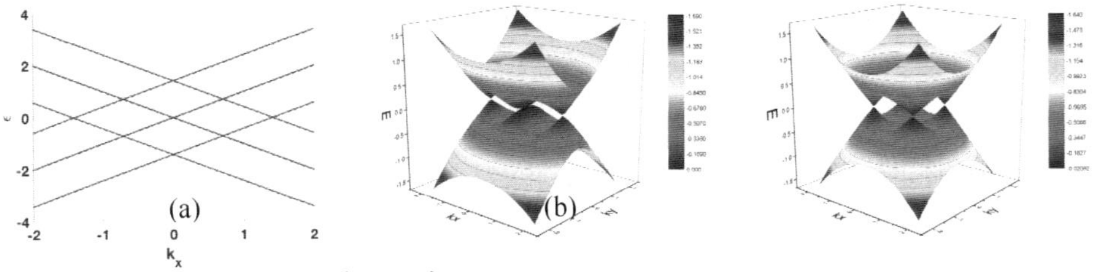

Figure. 4: Electron spectra $\varepsilon(k_x, k_y)$ for graphene subjected to linearly polarised laser fields with a static spatially-modulated periodic electric field. In the calculations we used the parameters $\omega = 1$, $\mu = 1$, $U_0 = 1$ and $A_0 = 1$ for each of chosen orientations, θ. (a) The spectrum shows gapless multiple crossing at the plane crossing the k_x axes. (b) showing two Weyl-Dirac located symmetrically with respect to the central Dirac point at K or K' place in the Brillouin Zone. All these three gapless points are oriented parallel to k_x at the value, $k_y = 0$. (c) The spectrum rotated by an angle equal to the laser's beam orientation, i.e., $\pi/2$, that is showing a broken special inversion symmetry and therewith forming two pairs of Weyl-Dirac cones symmetrically positioned around K or K' points of the Brillouin Zone.

To investigate in a detail the change in the spectrum, surface contour plots were shown at different laser field angles. For $\theta = 0$ there are three gapless points where the top and bottom energy bands touch each other. Near two peripheral points, the energy spectrum can be approximate as squeezed Dirac cones, $\varepsilon^2 = a\,(k_x)^2 + b(k_y)^2$, while near the middle point, the structure reminds a "cut" with energy $\varepsilon^2 = a(k_x)^2 + b(k_y)^4$, where a and b some real parameters. As the laser is rotating, two changes occur: 1)- the spectrum is rotating in a similar manner and 2)- the middle point experience a splitting where the energy cut transforms into two squeezed cones, see, the Figure 5-b, where the laser is applied at $\theta = \pi/4$. There spectrum has 4 Dirac points. When the laser field orientation changes, at the perpendicular case, when $\theta = \pi/2$, the spectrum is rotated by $\pi/2$ forming 4 Dirac-Weyl cones symmetric about the pole (see, the Figure 5-c), which is associated with K or K' of the graphene Brulloen Zone.

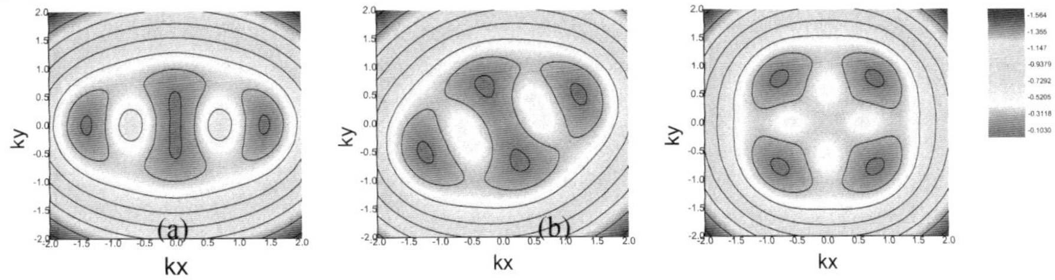

Figure. 5: The surface contour plot of the energy-momentum dispersion, $\varepsilon(k_x, k_y)$, for a monolayer graphene subjected to in-plane static periodically modulated electric field and a linear polarised laser field tilted at the angle θ to the graphene plane: (a) the case when the polarisation of the electromagnetic laser field coincides with the direction of the modulated electrical field, i.e. $\theta = 0$; Here we see a pair of the Weyl points located far apart from each other. In a centre there is an

elongated valley where a new pair of Weyl points is going to be nucleated, when the tilting angle increases, i.e. when $\theta \neq 0$; (b) The snapshot of the spectrum taken when the tilting angle $\theta = \pi/4$; We see already two pairs of the Weyl points here (c) The snapshot spectrum displaying two symmetrical pairs of the Weyl points taken at their perpendicular orientation, i.e. at $\theta = \pi/2$.

5. Conclusion

Here we shown that static periodically modulated magnetic/electric field together with linearly polarised laser field induce dynamical topological phase transitions in a single graphene layer. There in the vicinity of the original K and K' Dirac points the pairs of Weyl-Dirac cones are created. Then the graphene layer may be viewed as two-dimensional Weyl metal. The created Weyl metal energy spectrum is invariant under the laser field rotation when there is only spatial modulation of magnetic field. However, when the static electric field is periodically modulated, the system shows additional symmetry breaking phenomenon where additional pairs of the Weyl-Dirac cones are created. We show that this effect is controlled by a laser field tilting from the graphene plane where the modulated electric field is embedded. We expect that the described effect may be used in novel optoelectronic devices and the predicted phenomena will be observed in the future experiments.

References

[1] Novoselov, K S ; Geim, A K et al (2004), Science, 306(5696), 666-669.
[2] Meyer, J C ; Geim, A K ; et al (2007). Nature, 446, 60.
[3] O'hare, A., Kusmartsev, F. V., & Kugel, K. I. (2012). Nano letters, 12(2), 1045-1052.
[4] Geim , A. K. and Novoselov, K. S. (2007) The rise of graphene, *Nature Mater* 6, 183.
[5] Wallace, P. R (1947) Band theory of graphite, *Phys. Rev.* 71, 622.
[6] Yung, K. C., Wu, W. M., Pierpoint, M., & Kusmartsev, F. V. (2013). *Contemp. Phys.*, *54*, 233.
[7] Kusmartsev, F. V. and A. M. Tsvelik,(1985) JETP Lett., 42, 257-260.
[8] Pei, S., & Cheng, H. M. (2012). The reduction of graphene oxide. Carbon, 50(9), 3210-3228
[9] Georgakilas, V et al, (2012). Chemical Reviews, 112, 6156-6214.
[10] Bai, J. et al, (2010). Graphene nanomesh. Nature nanotechnology, 5(3), 190-194.
[11] Fang, T., et al. (2008). Physical Review B, 78(20), 205403.
[12] Rozhkov, A. V., Savel'ev, S., & Nori, F. (2009). Physical Review B, 79(12), 125420.
[13] Quhe, et al, (2012). NPG Asia Materials, 4, e6.
[14] Lu, T. M.,(2016). Scientific reports, *6. 20967.*
[15] Choi, S. M., Jhi, S. H., & Son, Y. W. (2010). Physical Review B, 81(8), 081407.
[16] Bliokh, Y. P., Freilikher, V., Savel'ev, S., & Nori, F. (2009). Physical Review B, 79(7), 07512
[17] Pedersen, T. G., et al. (2008). Physical Review Letters, 100(13), 136804
[18] Song, J. C., Shytov, A. V., & Levitov, L. S. (2013). Physical review letters, 111(26), 266801
[19] Alekseev, K. N., et al, (1998). Physica D: Nonlinear Phenomena, 113(2-4), 129-133
[20] Alekseev, K. N., et al . (1998). Phys. Rev. Lett., *80*(12), 2669.
[21] Alexeeva, N.; et al.,(2012) Phys. Rev. Lett., 109, 024102.
[22] Hramov, A. E., et al (2014). Phys. Rev. Lett., *112*(11), 116603.
[23] Savel'ev, S. E., & Alexandrov, A. S. (2011). Physical Review B, 84(3), 035428
[24] Rodriguez-Lopez, P., Betouras, J. J., & Savel'ev, S. E. (2014). Dirac fermion time-Floquet crystal: manipulating Dirac points. Physical Review B, 89(15), 155132
[25] Xia, F., Farmer, D. B., Lin, Y. M., & Avouris, P. (2010).Nano letters, 10(2), 715-718
[26] Zhang, Y., et al (2009). Nature, 459(7248), pp.820-823.
[27] Oostinga, J. B.,et al. *Nature Materials* 7(2), 151 – 157.
[28] Trabelsi, A. B. G.,et al. (2017). Nanoscale, 9, 11463-11474.
[29] Liu Yang, et al. (2015). Phys. Rev. B 92, 235438.
[30] R. D. Y. Hills, A. Kusmartseva, and F. V. Kusmartsev, (2017). , Phys. Rev. B 95, 214103.
[31] A. Kusmartseva and F. Kusmartsev (2017). Topological Materials: from polyacetylene to topological insulators and Weyl semimetals, Eur. Phys. J. B:-Conden.Matter Complex Syst. 90, 0001.

IOP Conf. Series: Journal of Physics: Conf. Series **961** (2017) 012013 doi:10.1088/1742-6596/961/1/012013

Superiority of localized surface plasmon resonance technique in characterization of ultra-thin metallic films

Sudheer[1,2,a], P. Tiwari[1,2], S. Bhartiya[2], C. Mukherjee[1,2], S. K. Rai[1,2], V. N. Rai[1,2,*], and A. K. Srivastava[1,2]

[1]Homi Bhabha National Institute, Mumbai, Maharashtra 400094, India
[2]Raja Ramanna Centre for Advanced Technology, Indore, Madhya Pradesh 452013, India

[a]E-mail: sudheer.rrcat@gmail.com, sudheer@rrcat.gov.in,
[*]Superannuated.

Abstract. The comparison and correlation of morphological, optical and crystallographic properties of ultra-thin Au films obtained using field-emission scanning electron microscopy (FESEM), x-ray reflectivity (XRR), UV-visible transmission, and grazing incidence x-ray diffraction (GIXRD) are presented. The Au thin films of different thickness are grown on the glass substrate using the sputtering technique. The particle size, number density and the covered area fraction of Au thin film are obtained from FESEM images. The XRR technique is used to determine the film thickness and surface roughness. The localized surface plasmon resonance (LSPR) response of Au thin films is obtained using UV-Vis transmission spectroscopy. The LSPR peak position and its strength are correlated with film morphology and thickness. Finally, it is shown that LSPR based spectroscopy techniques can provide much better information about morphology and thickness of the Au films up to a resolution of ~1 nm.

1. Introduction

The ultra-thin nanostructured metallic films of <10 nm thickness are of great interest because of their potential applications in nano-optics, energy harvesters like thin film solar cells and surface-enhanced Raman spectroscopy (SERS) [1,2,3,4]. The controlled fabrication of such nanostructured thin films along with accurate and fast characterization is very much crucial for its useful applications. Presently, among the various available techniques, fabrication of metallic thin film by the sputtering method has become popular due to its excellent control on the growth process [5]. The field-emission scanning electron microscopy (FESEM), and x-ray reflectivity (XRR) techniques are frequently used as the characterization tools for thin films. Amongst them, in FESEM, the charging effect and high vacuum requirement limit its applicability to conducting samples. The XRR technique provides accurate information about the thickness, surface and interface roughness, average mass density and electron density profile (EDP) of the continuous films [6,7]. But this technique did not provide significant information about the island and percolated type of films due to the absence of Keissing oscillations in the spectra. The localized surface plasmon resonance (LSPR) based spectroscopic technique that is the coherent oscillation of conduction band electron in nanostructures, is emerging as one of the superior tools especially for characterization of ultra-thin metal films (island and percolated) [8]. In the LSPR technique, optical transmission is dependent on particle size, shape, number density and the covered area fraction which can be further quantified for the characterization of ultra-thin films [9,10,11].

This article presents the comparison and correlation among various thin film characterization techniques. The LSPR based optical transmission is correlated with morphology and thickness of the film and found superior for characterization of the sputtered grown island and percolated films up to sub-nanometer resolution.

2. Experimental

The Au thin films are grown by a sputtering technique. The sputtering time was 10, 15, 20, 30, 45, and 60s at 10 mA current. The deposition was carried out on the glass substrate. Before the deposition process, a thorough cleaning of glass substrates was performed in a soap solution, acetone, and DI water. The characterizations of deposited films were carried out using FESEM, XRR, and UV-Vis transmission spectroscopy, and grazing-incident x-ray diffraction techniques (GIXRD). The ImageJ software is used to analyze the FESEM images. The Parratt formulation is applied for fitting the experimentally obtained XRR spectra [6].

3. Results and Discussion

3.1. *Morphological Characterization*

3.1.1. Imaging using scanning electron microscope

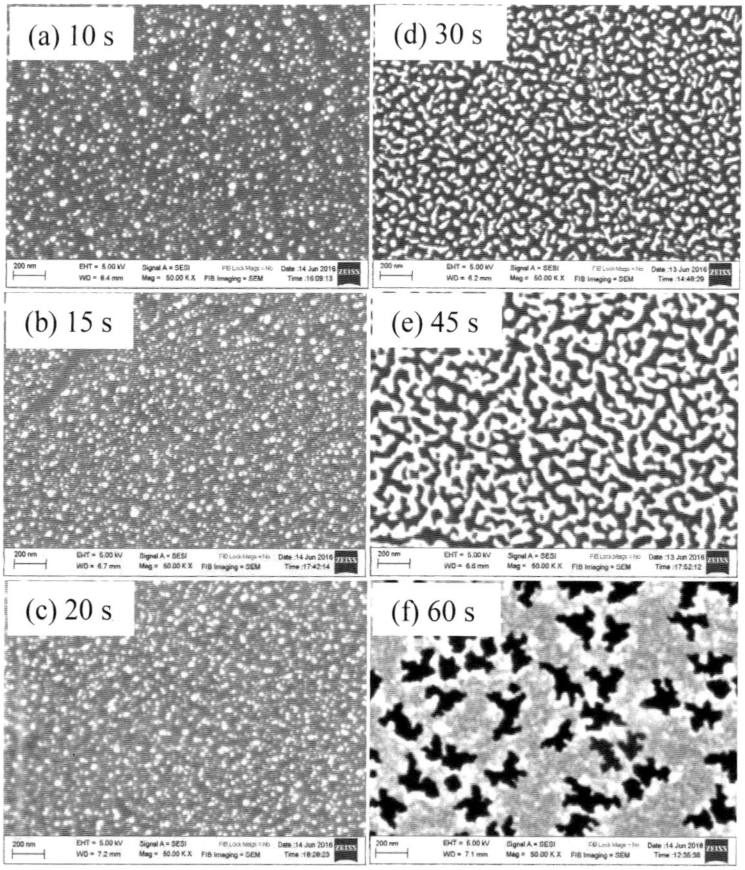

Figure. 1 Field-emission electron microscope image of Au thin films

First of all, the morphology of the fabricated Au thin films is imaged by FESEM. Figs. 1(a)-1(f) show the two-dimensional FESEM images of Au thin films of 10, 15, 20, 30, 45, and 60s samples. The quantitative analysis of FESEM images is performed in terms of average particle size (d) (Feret

Table-1: Variation in particle size, number density, covered area fraction and thickness of Au thin film.

Deposition time (s)	Particle size (nm)	Number density (particles /μm^{-2})	Covered area fraction (F%)	Thickness (nm)
10	24	3.74×10^{2}	13.66	1.4
15	25	6.55×10^{2}	19.31	2.1
20	29	5.29×10^{2}	23.99	2.8
30	72	1.62×10^{2}	33.63	4.2
45			52.45	6.3
60			72.75	8.4

diameter), number density (n), and covered area fraction (F%) [Table-1]. The Feret diameter is defined as $d=P/\pi$, where P is the perimeter of Au nanoparticle [12]. Figs. 1(a)-1(b) show the average particle size is almost same, and shape is nearly spherical for 10 and 15s films while the particle number density in the 15s film is almost twice in comparison to the 10s film (Table-1). For the 20s film, particle size increases while number density decreases. It is widely reported in the literature that both of the above facts probably occurs only due to coalescence phenomenon. In which the nearest neighbor small particles come in the contact and form bigger particles. In the case of 30s film average particle sizes abruptly increases. In which the small sizes particle have a nearly spherical shape while larger size particles have complex chain-like structures. The few bigger chains structure has sizes ~280 nm also exist. In case of the 45s sample, size of the complex chain structures are further increases, and interconnected networks are formed in the film. Such kind of films are well known as percolated films. After that in the case of 60s sample, the percolated structures finally transformed in the continuous patches of films (semicontinuous film). The covered area fraction (F%) for 10s sample is found ~14 %. It to be noted that F% consistently increases with the deposition time and reaches up to ~73 % for 60s sample.

3.1.2. X-ray reflectivity

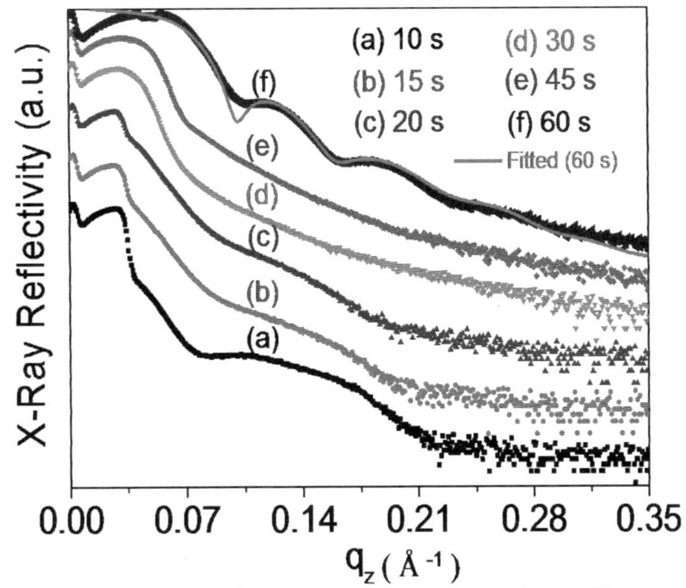

Fig. 2 X-ray reflectivity spectra of sputter grown Au thin films

Fig. 2 shows the XRR spectra of Au thin film deposited for different times. For 60s sample, well defined Keissing oscillations are found in the XRR spectra [7]. The experimental data of 60s sample is fitted with simulated data. The thickness and roughness values are estimated as 8.4 nm and 1.3 nm respectively. As described above that 60s sample shows the presence of large patches of continuous film in the FESEM images (Fig. 1(f)). These patches are supportive in XRR measurement and produce oscillation in the recorded spectra. These oscillations damped for all other films (10-45s) due to an island or percolated structures (Figs. 1(a)-1(e)). The island and percolated structure reduces the specular reflection and enhances the scattering of the incident x-ray beam that results in vanished Keissing oscillations in the spectra. Therefore, XRR technique is limited for semicontinuous and continuous films. Only a small information can be extracted for island or percolated films from the XRR technique. Thus, for all other lower thickness films, average mass-thickness is determined by scaling with time and shown in Table-1[12,13].

3.2. Optical characterization

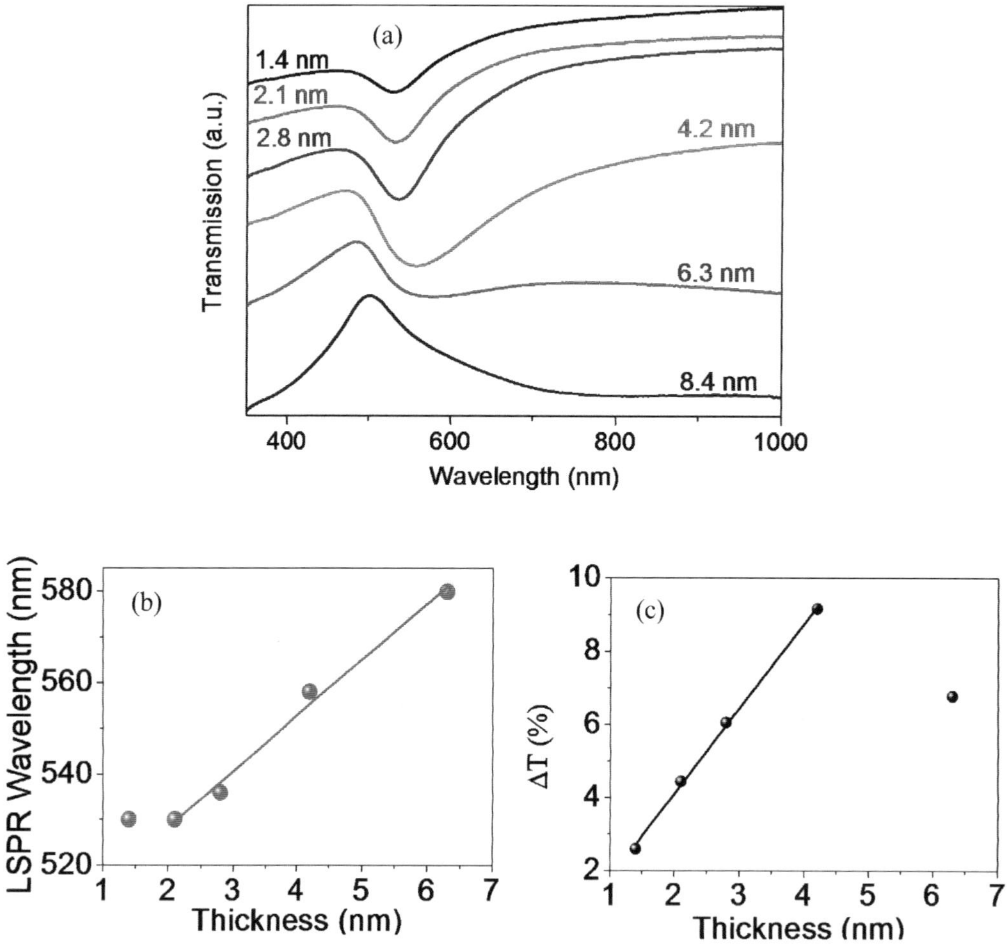

Fig. 3 (a) UV-Vis transmission spectra of Au thin films, (b) variation in LSPR dip position with film thickness and (c) variation in LSPR dip strength with film thickness.

To verify the above scaling rule of thickness and to explore the superiority of LSPR based technique, the transmission spectra of the Au thin films are recorded. Fig. 3(a) shows the transmission spectra of sputtered grown Au thin films. A dip at ~530 nm wavelength in the transmission spectra is found for

IOP Conf. Series: Journal of Physics: Conf. Series **961** (2017) 012013 doi:10.1088/1742-6596/961/1/012013

1.4 nm thickness film. This fall in the transmission spectra is due to the presence of random Au islands in the films showing excitation of characteristic LSPR. The LSPR minima position and its broadening are intimately dependent on size and shape of the nanoparticles as well as their decoration on the substrate. A similar LSPR peak position in the 1.4 nm and 2.1 nm thickness film is the clear evidence of the existence of the same particle size. A variation in particle number density is strongly witnessed to increase the LSPR strength in 2.1 nm thickness films. Moreover, a small change in the participle size distribution can produce inhomogeneous broadening in the LSPR profile which can be realized by variation in its full width at half maxima (FWHM). The increment in the particle size leads to a consistent redshift of LSPR peak position of the film as thickness increases from 2.1 to 4.2 nm (Fig. 3(a)). The 6.3 nm thickness film shows the convoluted optical response of both island and percolated film. Finally, in the case of 8.4 nm thickness film, spectra become like a typical optical response of highly percolated films. Furthermore, the optical response of Au thin films is quantified in terms of peak position (λ), and its strength ($\Delta T\%$) according to the methodology described in the literature [10,11,13]. Figs. 3(b)-3(c) show the variation in λ and $\Delta T\%$ with thickness. All the experimentally obtained data are fitted linearly with slopes 12.3 $\pm$0.8 and 2.3$\pm$ 0.1 % respectively. It is well known that most of the spectrophotometer has wavelength and transmittance resolution of ~1 nm and 0.1 % respectively. Which clearly indicated that thickness of the ultra-thin metal films can be obtained with the resolution of sub-nanometer scale.

3.3. Crystallographic characterization

The GIXRD spectra were also recorded for 1.4, 2.8, 4.2 and 8.4 nm thickness Au thin films. The peaks in the XRD spectra are indexed as (111), (200), (220) and (311) planes of face-centered cubic (FCC) Au crystal structures[14]. The spectra indicate that for <4.2 nm thickness film GIXRD technique could also not deliver the significant information about the crystal structure and grain size present in the film.

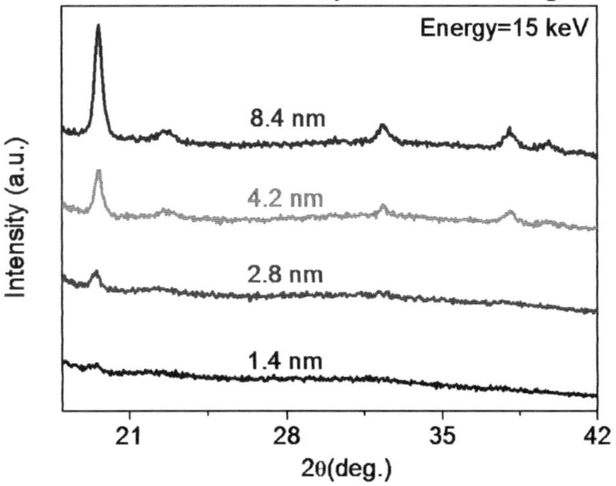

Fig. 4 X-ray diffraction spectra of Au thin films

4. Conclusions

The sputter grown Au thin films of thickness <10 nm have been characterized using various techniques. The morphology and thickness of the film are obtained by FESEM and XRR techniques. The localized surface plasmon resonance (LSPR) response is measured using UV-Vis optical transmission, and their results are quantified and correlated with the morphology and thickness. The correlation suggests that LSPR technique can be used to find the film thickness up to sub-nanometer resolution. Finally, LSPR based characterization is found to be a superior technique for characterization of ultra-thin Au films.

Acknowledgment:
One of the authors Sudheer acknowledges the Homi Bhabha National Institute (HBNI) for providing financial assistance. Authors are thankful to Mr. Mahendra Babu for providing help during the deposition of Au thin films. Authors are also thanked Mr. Rajiv Kamparath for providing active guidance during optical measurements.

References

[1] K. Wen, Y. Hu, L. Chen, J. Zhou, L. Lei and Z. Guo, Opt. Comm. **336**, 243 (2015).
[2] Y. A. Akimov, W. S. Koh, K. Ostrikov, Opt. Exp. **17**, 10195 (2009).
[3] A. Merlen, V. Gadenne, J. Romann, V. Chevallier, L. Patrone, J. C. Valmalette, Nanotechnology **20**, 215705 (2009).
[4] G. Q. Liu, Z. Q. Liu, Y. H. Chen, K. Huang, L. Li, F. L. Tang, L. X. Gong, Y. Hu, X. N. Zhang, Optik **124**, 5124 (2013).
[5] T. D. Phuc, M. Yoshine, A. Yamanaka, and T. Yamamoto, Int. J. of Automation Technology **8**, 1 (2013).
[6] L. G. Parratt and C. F. Hempstead, Phy Rev. Lett. **94**, 1593 (1594).
[7] I. Stabrawa, D. Banas, K. Dworecki, A. Kubala-Kukus, J. Braziewicz, U. Majewska, J. Wudarczyk-Mocko, M. Pajek and S. Góźdź, ACTA PHYSICA POLONICA A **129**, 233 (2016).
[8] A. Herz, A. Franz, K. Theska, M. Hentschel, Th. Kups, D. Wang, and P. Schaaf, AIP Advances **6**, 035109 (2016).
[9] U. Kreibig and M. Vollmer, Optical Properties of Metal Clusters (Springer, Heidelberg, 1995).
[10] Gong, Y. Zhou, L. He, B. Xie, F. Song, M. Han and G. Wang, Eur. Phys. J. D **67**: 87 (2013).
[11] S. Verm, B. T. Rao, S. Bhartiya, V. Sathe, L. M. Kukreja, App. Surf. Sci. **346**, 379 (2015).
[12] Sudheer, Puspen Mondal, V. N. Rai, and A. K. Srivastava, AIP Advances **7**, 075303 (2017).
[13] S. Verma, B. T. Rao, S. Rai, V. Ganesan, and L. M. Kukreja, Appl. Sur. Sci. **258**, 4898 (2012).
[14] V. Svorcík, O. Kvítek, J. Ríha, Z. Kolská, J. Siegel, Vacuum **86**, 729 (2012).

IOP Conf. Series: Journal of Physics: Conf. Series **961** (2017) 012014 doi:10.1088/1742-6596/961/1/012014

Silicon Oxycarbide Waveguides for Photonic Applications

Faisal Ahmed Memon[1,2,3], Francesco Morichetti[1,2] and Andrea Melloni[1,2]

[1] Dipartimento di Elettronica, Informazione e Bioingegneria (DEIB), Politecnico di Milano, via Ponzio 34/5, 20133 Milan Italy
[2] Polifab, Micro-nano Technology Center, Politecnico di Milano, Via Giuseppe Colombo, 81, 20133 Milan Italy
[3] Department of Telecommunications Engineering, Mehran University of Engineering & Technology Jamshoro 76062 Pakistan

faisalahmed.memon@polimi.it, faisal.memon@faculty.muet.edu.pk

Abstract. Silicon oxycarbide thin films deposited with rf reactive magnetron sputtering a SiC target are exploited to demonstrate photonic waveguides with a high refractive index of 1.82 yielding an index contrast of 18% with silica glass. The propagation losses of the photonic waveguides are measured at the telecom wavelength of 1.55 μm by cut-back technique. The results demonstrate the potential of silicon oxycarbide for photonic applications.

1. Introduction

Integrated photonics has progressed significantly over the past decades. Photonics is capable of providing reliable and low-cost high-speed communications. Though different materials have been exploited to fulfil the needs of the integrated photonics industry, however the well-established platforms such as SiO_2, SiN, Si, and InP are limited in their refractive index tunability. Silicon oxynitride has been demonstrated to show the flexibility of refractive index n tuning between silica glass and SiN. First, SiON exhibits absorption peak around 1500 nm, and secondly cracks are observed when refractive index becomes larger than 1.5 that confines its use to low index [1]. Silicon oxycarbide (SiOC), on the other hand, has a possibility of wide refractive index tunabilty ranging from silica glass to amorphous silicon carbide [2]. The tunability in refractive index is useful for the large integration scale of photonic devices on a single chip. The light mode can be confined tightly in high contrast waveguide core that enables smaller bending radii and reduces device footprint [3]. The overall picture of the state-of-the art of optical materials with SiOC being a promising platform is represented by Figure 1.

Silicon oxycarbide has received considerable attention in the scientific community. Showing material versatility, SiOC is applied in various important applications ranging from photoluminescence, multilayer structures for bloch surface waves to inter-layer dieclectric. Despite possessing excellent material properties, silicon oxycarbide is yet not established as a platform for integrated photonics.

In this work, silicon oxycarbide thin films are deposited with rf reactive magnetron sputter technique. The thin films are characterized with several tools to better understand the structure and optical properties. Then the films of SiOC are exploited as core layer material to demonstrate waveguides for integrated photonics with high refractive index of 1.82 that is surrounded by silica glass cladding. The losses of the SiOC waveguides were estimated by butt-coupling with optical fibers using a controlled optical setup.

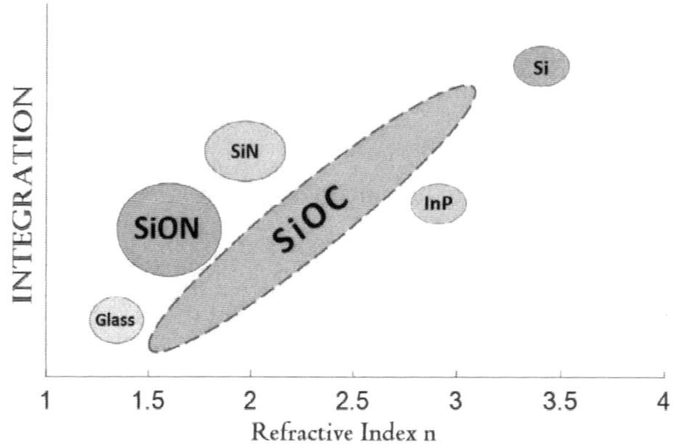

Figure 1. State-of-the-art of photonic integration technology with silicon oxycarbide (SiOC) as a promising platform.

2. Experimental Details

Silicon oxycarbide (SiOC) thin films were produced by sputtering a 99.9% pure SiC target in a rf reactive magnetron sputter chamber in the presence of argon (Ar) and oxygen (O_2) gases. The SiC target material placed on cathode was negatively biased by suppling rf power and the plasma (Ar) and reactive (O_2) gases were inlet in the chamber and controlled with mass flow controllers. SiOC thin films with varying refractive index were produced by increasing sputtering power from 150 to 450 watts while keeping Ar and O_2 gases at a constant ratio. The SiOC films were deposited on Si (100) and Si (100) with SiO_2 layer substrates. The SiOC films deposited on Si substrates with SiO_2 layer were used to fabricate photonic channel waveguides.

The deposited SiOC films were characterized to investigate the effect of sputtering parameters and attain the favourable properties. Surface roughness being one of the important parameter was quantified with atomic force microscope (5600LS Keysight AFM system) by scanning the film in tapping mode. The cross section of SiOC waveguides and films were analysed with scanning electron microscopy (SEM-LEO 1525) for morphology and determine the thickness. The optical properties (index of refraction n and coefficient of extinction k) of SiOC films were studied with variable angle spectroscopic ellipsometer (Woollam Inc., US) from UV to near-infrared spectral range.

The photonic waveguides in SiOC films with $n = 1.82$ were fabricated by UV photo-lithography and reaction of ion etching (R.I.E) techniques. The fabricated waveguides were covered by a thicker PECVD SiO_2 ($n = 1.444$) top layer to attain symmetric guided modes. The SiOC waveguides were characterized for propagation losses with small-core fibers (MFD = 3.6 μm) on a polarization controlled optical bench.

3. Results and Discussion

The surface roughness of the deposited SiOC films was quantified with AFM by scanning in tapping mode over a surface area of 100 μm^2. The rms roughness as low as 0.9 nm was evaluated from AFM measurements for a film thickness of 400 nm, that is atomically flat to realize photonic waveguides. The film surface roughness is important as it acts as source of scattering for propagating light along the waveguide and contributes to the losses.

IOP Conf. Series: Journal of Physics: Conf. Series **961** (2017) 012014 doi:10.1088/1742-6596/961/1/012014

The optical properties (n and k spectra) of SiOC thin films deposited under varying sputtering conditions are shown in Figure 2. The refractive index n increases as rf power gains from 150 to 450 watts. The increase in n indicates change in the chemical composition of SiOC films. The n of SiOC ranges from 1.45 to ~1.82 demonstrating the wider tunability as shown in Figure 2(a). The extinction coefficient k of SiOC thin films shown in Figure 2(b) is negligible, that is below the detection limit of ellipsometer in the near-infrared spectrum. The lower material absorption coefficient enables the fabrication of photonic waveguides in SiOC thin films. The tunable refractive index and low absorption are ideal for the realization of a range of photonic devices in a single material platform [4].

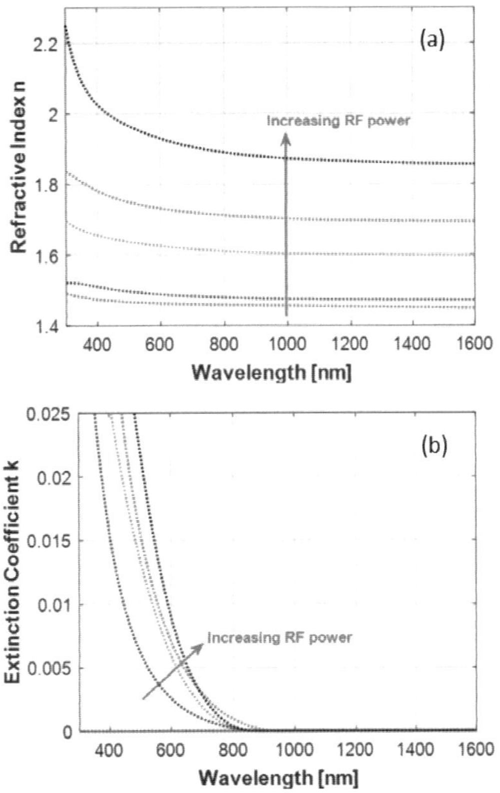

Figure 2. (a) Index of refraction n (b) coefficient of extinction k spectra of SiOC films deposited under changing sputtering conditions.

Figure 3 shows the photonic waveguide fabricated with SiOC core. The image in Figre 3 was captured with SEM. The SiOC waveguide thickness was around 400 nm and refractive index n = 1.82. To achieve single mode operation, the width of SiOC waveguides is 1.8 μm that is patterned with UV photo-lithography and reactive ion etching as shown in Figure 3.

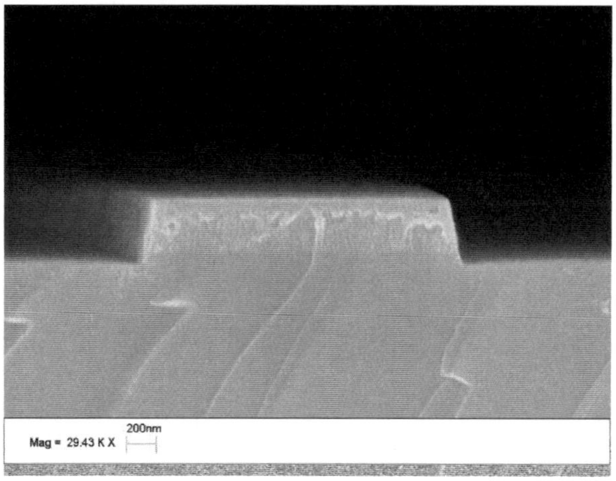

Figure 3. SiOC photonic waveguide with width = 1.8 μm, height = 400 nm and n = 1.82.

The propagation loss of the SiOC core waveguide were characterized on a polarization controlled optical bench. The losses of propagation as low as 0.8 dB/mm were measured for SiOC waveguides with n = 1.82 for both TE and TM polarizations.

4. Conclusion

SiOC photonic waveguides with high refractive index of 1.82 are demonstrated for the first time. The propagation losses of the waveguides are characterized and presented. The results show the significance of SiOC material and imply its suitability for photonic integration applications.

References

[1] Wörhoff K, Driessen A, Lambeck P V, Hilderink L T H., Linders P W C and Popma Th J A 1999 *Sen. Act. A: Phys.* **74** 9–12.
[2] Ryan J V and Pantano C G 2007 *J. Vac. Sc. & Technology A* **25** 153-9.
[3] Melloni A, Costa R, Cusmai G and Morichetti F 2009 *Int. J. Materials and Product Technology* **34** (4).
[4] Memon F A, Morichetti F, Abro M I, Iseni G, Somaschini C, Aftab U and Melloni A 2017 *EPJ Web Conf.* **139**.
[5] Memon F A, Morichetti F and Melloni A 2017 *Appl. Sci.* **7** 561.